Vector Optimization

Series Editor:

Johannes Jahn
University of Erlangen-Nürnberg
Department of Mathematics
Cauerstr. 11
81058 Erlangen
Germany
johannes.jahn@fau.de

For further volumes:
http://www.springer.com/series/8175

Vector Optimization

The series in Vector Optimization contains publications in various fields of optimization with vector-valued objective functions, such as multiobjective optimization, multi criteria decision making, set optimization, vector-valued game theory and border areas to financial mathematics, biosystems, semidefinite programming and multiobjective control theory. Studies of continuous, discrete, combinatorial and stochastic multiobjective models in interesting fields of operations research are also included. The series covers mathematical theory, methods and applications in economics and engineering. These publications being written in English are primarily monographs and multiple author works containing current advances in these fields.

Gabriele Eichfelder

Variable Ordering Structures in Vector Optimization

 Springer

Gabriele Eichfelder
Institut für Mathematik
Technische Universität Ilmenau
Ilmenau
Germany

ISSN 1867-8971   ISSN 1867-898X (electronic)
ISBN 978-3-662-52551-7   ISBN 978-3-642-54283-1 (eBook)
DOI 10.1007/978-3-642-54283-1
Springer Heidelberg New York Dordrecht London

Printed on acid-free paper

Springer is part of Springer Science+Business Media (www.springer.com)

To Tom.

Preface

In mathematical optimization one aims at minimizing or maximizing an objective function over some feasible set. In case the objective function is scalar-valued it is straightforward how to define an optimal solution: a feasible element with the smallest (or largest) scalar as objective function value is an optimal solution.

However, in case the objective function is vector-valued, i.e. it maps in a real linear space, it is not obvious anymore how to compare the values of the objective function. In the applied sciences Edgeworth [38] and Pareto [127] were probably the first who introduced an optimality concept for vector optimization problems, i.e., for optimization problems with such a vector-valued objective function, cf. [53, 113]. Both have studied the so-called *multiobjective optimization* problems which are vector optimization problems with an objective function mapping in the m-dimensional Euclidean space $\mathbb{R}^m$ for $m \geq 2$. For comparing elements in $\mathbb{R}^m$ they used the natural ordering cone, the nonnegative orthant, which corresponds to the componentwise partial ordering. Based on that, in vector optimization it is often assumed that the linear space is partially ordered by a convex cone and an element is an optimal element (referred to as an efficient element) of a set if it is not dominated by (i.e., worse than) any other reference element w.r.t. the partial ordering of the linear space.

But already in the first publications in mathematics in the 1970s related to the definition of optimal elements in vector optimization [19, 158] also the idea of variable ordering structures was given: it was assumed that there is a set-valued map with cone values that associates with each element of the linear space and ordering. A candidate element is called a nondominated element if it is not dominated by other reference elements w.r.t. the corresponding ordering of these other ones. In addition to the notion of nondominated elements, in [28–30] they also consider another notion of optimal elements in case of a variable ordering structure. Namely, a candidate element is called a minimal element (also called nondominated-like) if it is not dominated by any other reference element w.r.t. the ordering of the candidate element.

This book aims at providing an introduction to vector optimization with a variable ordering. Application problems are presented which motivate the study of these optimization problems. Next to a comprehensive basic theory, numerical approaches are discussed which allow to solve such problems in practice. Throughout this book we assume that the variable ordering structure is defined in the objective space by a set-valued map which associates with each element of the space a cone of preferred or dominated directions.

The recent interest in vector optimization problems with a variable ordering structure started some years ago with an application problem in image registration in medical engineering [146]: there it is the aim to merge several medical images gained by different imaging methods as, for instance, computer tomography, magnetic resonance tomography, positron emission tomography, or ultrasound. One searches for a best transformation map, also called registration. To measure the quality of such a transformation, a multitude of similarity measures is known which all possess different properties and advantages. The values of the different measures can be interpreted as objectives which have to be minimized all at the same time.

However, depending on the objective values, it is advantageous to put a higher weight on some of these objectives than on others. To each element in the objective space one can thus relate a weight vector and consider a related cone which depends on the weight vector and hence on the considered element. Then one searches for an optimal solution of a vector optimization problem with a variable ordering structure. We discuss this application in more detail in Sect. 1.3.1.

Already in [99] Karaskal and Michalowski recognized that the importance of criteria may change during the decision-making process and that it may depend on the current objective function values. Wiecek gives in [154] an example for that. In [9] Baatar and Wiecek examine the concept of equitability, which has applications in portfolio optimization and location problems. They show that this minimality notion is related to a finite number of ordering cones or polyhedral sets instead of a unique ordering cone. The objective space is partitioned into sections and the related ordering is variable and depends on the section in which an element lies. More details are given in Sect. 1.3.2.

Engau examines in [58] the role of variable ordering structures in preference modeling. He gives examples showing the limitations of preference modeling using only one ordering cone. Variable structures defined by convex cones containing the nonnegative orthant and being some kind of symmetric are studied. These convex cones are special Bishop-Phelps cones (BP cones) and BP cones play an important role for some scalarization results in Chap. 6.

We start in Chap. 1 by recalling basic concepts of partially ordered spaces and basic results on ordering cones and dual cones. Then we introduce variable ordering structures. We assume that the variable ordering structure is defined by a set-valued map called ordering map. Two binary relations defined by such an ordering map are introduced and their properties are examined. We also focus on special ordering maps where the images are BP cones because such ordering maps are important in applications and also for theoretical results. Moreover, we discuss the applications mentioned above which illustrate that partial orderings are not always an adequate tool for modeling real-world problems.

In Chap. 2 we collect several optimality notions based on the two binary relations introduced in Chap. 1. We compare these new concepts with known concepts in partially ordered spaces, and we show that the main two different optimality concepts w.r.t. a variable ordering structure, the minimal and the nondominated elements, are in general not related in the sense that one does not imply the other. We continue the chapter by providing first results on characterizations of the optimal elements.

The variable ordering structures are defined by ordering maps which are cone-valued maps. For that reason we study in Chap. 3 cone-valued maps. We examine classical properties, formerly introduced for arbitrary set-valued maps, like convexity, cone-convexity, linearity, or monotonicity. It turns out that some of these properties like convexity directly imply that the cone-valued map is constant. This is important to know for the study of scalarization functionals. Convexity of the ordering map would imply convexity of the scalarization functionals studied in Chap. 5 but is thus a too strong assumption. In case of non-appropriateness of the classical notions we propose new concepts.

In Chaps. 4–6 linear and nonlinear scalarization functionals are proposed and their properties are studied. With these functionals at hand a vector optimization problem can be replaced by a scalar-valued optimization problem which allows, for instance, the formulation of optimality conditions of Fermat and Lagrange type or can be used as the base of numerical solution methods. In Chap. 4 we start with linear scalarization functionals which turn out to be appropriate in case of convexity of the considered set only. But also under convexity assumptions the necessary conditions for nondominated elements are in general too weak and the sufficient conditions are too strong, and a complete characterization of the optimal elements is not possible.

For that reason we discuss nonlinear scalarization functionals in Chap. 5 which allow a complete characterization of nondominated and minimal elements. We consider a modification of the so-called signed distance functional which was introduced by Hiriart-Urruty, and of a second functional called translative functional, which generalizes a functional known in the literature as Gerstewitz or Tammer-Weidner functional or Pascoletti-Serafini scalarization. For both scalarization functionals it can in general not be assumed that they are convex in case of a variable ordering structure. However, convexity is required for the formulation of sufficient optimality conditions of Fermat and Lagrange type for the vector optimization problems.

This leads to Chap. 6 in which we concentrate on variable ordering structures which are defined by ordering maps with images being BP cones. This additional structure allows to introduce a new scalarization functional which is also new in partially ordered spaces. This functional allows a complete characterization of the nondominated and also the minimal elements and, what is more, is convex at least under strong assumptions.

We provide in Chap. 7 subdifferential information for the scalarization functionals introduced in Chap. 6. Then we are able to formulate necessary and sufficient optimality conditions of Fermat and Lagrange type for unconstrained and constrained vector optimization problems with (set-valued) objective maps mapping in a real linear space equipped with a variable ordering structure. For defining optimal solutions of an optimization problem with a set-valued objective map we choose here the vector approach, i.e., optimal solutions are defined as pre-images of optimal elements of the image set of the feasible set under the objective map.

These new scalarization functionals based on the structure of BP cones are also used in Chap. 8 for obtaining duality results, i.e., for defining a dual set to the original vector optimization problem with the optimal elements of the dual set being related to the optimal solutions of the original problem. We provide in Chap. 8 also duality results based on linear scalarizations as well as results concerning general duality for a primal and a dual set. It is interesting to see that the two optimality concepts, the nondominated and the minimal elements, which are in general not related in the sense that one does not imply the other, are related by duality results.

Chapter 9 gives a survey on numerical approaches for solving vector optimization problems with a variable ordering structure. We provide algorithms for solving finite discrete as well as continuous vector optimization problems without any significant restrictions.

In the final chapter we give a short outlook on the appearance of variable ordering structures in vector variational inequalities, vector complementarity, and equilibrium problems. We show that also the theory of consumer demand in economics is related to variable ordering structures. Finally we discuss an application in the treatment planning in intensity-modulated radiation therapy which shows that a cone-valued ordering map might in some applications be an adequate concept only locally which gives rise to future examinations.

I am very grateful to Professor J. Jahn for valuable discussions and hints. I am also grateful to Professor D.T. Luc, C. Gebhardt, M. Pruckner, and M. Ziegler for useful comments, as well as to Professor T.X.D. Ha, Professor R. Kasimbeyli, and Dr. T. Gerlach who worked with me on some topics in vector optimization with a variable ordering structure. Moreover, I am indebted to A. Eger and Dr. T. Gerlach for their support.

Ilmenau, Germany   Gabriele Eichfelder
December 2013

Contents

Chapter 1
Variable Ordering Structures

In scalar-valued optimization the definition of an optimal solution is quite natural: let S be some feasible set in a real linear space X and let $f \colon S \to \mathbb{R}$ be the objective function which should be minimized over S. Then $\bar{x} \in S$ is called an optimal solution of

$$\min_{x \in S} \; f(x)$$

if

$$f(\bar{x}) \leq f(x) \; \text{ for all } \; x \in S \,.$$

The natural ordering $\leq$ describes here a total ordering in $\mathbb{R}$, i.e. all elements in $\mathbb{R}$ are comparable using this ordering.

However, when we consider vector optimization problems, i.e. optimization problems with a vector-valued objective map $f \colon S \to Y$ mapping in some real linear space Y, then the ordering of the elements in Y is not as obvious as in the scalar-valued case. In general, not a total ordering is given but at most a partial ordering, which is a reflexive and transitive binary relation that is compatible with the linear structure of the space. Such a partial ordering can be represented by a convex cone, a so-called ordering cone.

Relations which are not defined by one single convex cone but by a family of convex cones, one for each element in Y, are denoted variable ordering structures. These families of cones describe a set-valued map which associates with each element in Y a convex cone in Y. Despite the name variable *ordering*, transitivity and compatibility with the linear structure of the space are generally not given.

In this chapter, we introduce variable ordering structures and examine their properties. We present special variable orderings given by Bishop-Phelps cones and we give a selection of applications of vector optimization problems which use a variable ordering structure.

G. Eichfelder, *Variable Ordering Structures in Vector Optimization*, Vector Optimization,
DOI 10.1007/978-3-642-54283-1_1, © Springer-Verlag Berlin Heidelberg 2014

1.1 Partial Orderings and Variable Ordering Structures

As mentioned above, for defining optimality for a vector optimization problem, we have to define first how to compare elements in a real linear space. A classical concept is the one of a partial ordering which is based on a binary relation on the space.

Definition 1.1. Let Y be a real linear space.

(a) A nonempty subset R of the product space $Y \times Y$ is called a *binary relation R* on Y. We write yRz for $(y, z) \in R$.

(b) A binary relation $\leq$ on Y is called a *partial ordering* on Y, if for arbitrary $w, x, y, z \in Y$

   (i) (reflexivity) $x \leq x$,
   (ii) (transitivity) $x \leq y, \ y \leq z \Rightarrow \ x \leq z$,
   (iii) $x \leq y, \ w \leq z \Rightarrow \ x + w \leq y + z$,
   (iv) $x \leq y, \ \alpha \in \mathbb{R}_+ \Rightarrow \ \alpha x \leq \alpha y$.

(c) A partial ordering $\leq$ on Y is called *antisymmetric*, if for arbitrary $y, z \in Y$

$$y \leq z, \ z \leq y \ \Rightarrow \ y = z.$$

A real linear space equipped with a partial ordering is called a *partially ordered linear space*. The axioms (iii) and (iv) guarantee the compatibility of the partial ordering with the linear structure of the space. For instance, the componentwise ordering in $\mathbb{R}^m$, defined for all $x, y \in \mathbb{R}^m$ by

$$x \leq y \ \Leftrightarrow \ x_i \leq y_i \ \text{ for all } i = 1, \ldots, m,$$

is an antisymmetric partial ordering.

Partial orderings are closely related to convex cones.

Definition 1.2 ([94, Definition 1.10]). Let Y be a real linear space.

(a) A nonempty set $K \subset Y$ is called a *cone*, if

$$y \in K, \ \lambda \geq 0 \Rightarrow \lambda y \in K.$$

(b) A cone K is called *pointed* if

$$K \cap (-K) = \{0_Y\}.$$

(c) A nonempty convex subset B of a convex cone $K \neq \{0_Y\}$ is called a *base* for K, if each $y \in K \setminus \{0_Y\}$ has a unique representation of the form

$$y = \lambda y \ \text{ for some } \lambda > 0 \text{ and some } b \in B.$$

(d) Let $\Omega \subset Y$ be a nonempty set. The cone

$$\text{cone}(\Omega) := \{\lambda\, y \in Y \mid \lambda \geq 0,\; y \in \Omega\}$$

is called the *cone generated by* Ω.

(e) Let $\Omega \subset Y$ be a nonempty set. The convex set

$$\text{conv}(\Omega) := \bigcap_{\substack{\Omega \subset M \\ M\,\text{convex}}} M$$

is called the *convex hull of* Ω.

Hence, for a base B of a convex cone K, it holds cone $(B) = K$. A cone $K \neq \{0_Y\}$ is said to be a *nontrivial cone*. It is easy to show that any nontrivial convex cone with a base is pointed.

Lemma 1.3 ([94, Lemma 1.11]). *Let Y be a real linear space. A cone $K \subset Y$ is convex if and only if*

$$K + K \subset K.$$

The following theorem gives the relation between partial orderings and convex cones:

Theorem 1.4 ([94, Theorem 1.18]). *Let Y be a real linear space.*

(a) If $\leq$ is a partial ordering on Y, then the set

$$K := \{y \in Y \mid y \geq 0_Y\}$$

is a convex cone. If, in addition, $\leq$ is antisymmetric, then K is pointed.

(b) If $K \subset Y$ is a convex cone, then the binary relation

$$\leq_K := \{(y, z) \in Y \times Y \mid z - y \in K\}$$

is a partial ordering on Y. If, in addition, K is pointed, then $\leq_K$ is antisymmetric.

Definition 1.5. A convex cone which characterizes a partial ordering on a real linear space is called an *ordering cone*.

For example, the sets $K_1 := \mathbb{R}_+^2$ and

$$K_2 := \{x \in \mathbb{R}^2 \mid 2x_1 + x_2 \leq 0,\; -0.5x_1 - x_2 \leq 0\}$$

are convex cones and may thus be ordering cones in $\mathbb{R}^2$, while the set $K_1 \cup K_2$ is a cone which is not convex. For $\leq$ the componentwise ordering in $\mathbb{R}^m$, the associated ordering cone is $\mathbb{R}_+^m$. The cones K_1, K_2 and $K_1 \cup K_2$ are all pointed cones.

In the following we also need the concepts of a dual cone and the quasi-interior of the dual cone.

Definition 1.6. Let Y be a real linear space with a convex cone K and let Y' denote the algebraic dual space.

(a) The cone

$$K' := \{y' \in Y' \mid y'(y) \geq 0 \text{ for all } y \in K\}$$

is called the (algebraic) *dual cone* for K.

(b) The set

$$K^\#_{Y'} := \{y' \in Y' \mid y'(y) > 0 \text{ for all } y \in K \setminus \{0_Y\}\}$$

is called the (algebraic) *quasi-interior of the dual cone* for K

For Y a real topological linear space and Y^* the topological dual space with the induced norm $\| \cdot \|_*$, we denote by

$$K^* = \{y^* \in Y^* \mid y^*(y) \geq 0 \text{ for all } y \in K\}$$

the (topological) dual cone and by

$$K^\# := K^\#_{Y^*} = \{y^* \in Y^* \mid y^*(y) > 0 \text{ for all } y \in K \setminus \{0_Y\}\}$$

the (topological) quasi-interior of the dual cone. For $Y = \mathbb{R}^m$ and K a convex cone in $\mathbb{R}^m$, the definitions of the dual cone and the quasi-interior of the dual cone read as

$$K^* = \{l \in \mathbb{R}^m \mid l^T y \geq 0 \text{ for all } y \in K\} \text{ and}$$
$$K^\# = \{l \in \mathbb{R}^m \mid l^T y > 0 \text{ for all } y \in K \setminus \{0_{\mathbb{R}^m}\}\} .$$

For $K = \mathbb{R}^m_+$ the dual cone is $K^* = \mathbb{R}^m_+$ and the quasi-interior of the dual cone is $K^\# = \{y \in \mathbb{R}^m \mid y_i > 0, \ i = 1, \ldots, m\}$.

The following lemma gives useful characterizations of the elements of a cone K by the elements of the dual cone. Recall that the *algebraic interior* of a nonempty set $\Omega \subset Y$ is the set

$$\text{cor}(\Omega) := \{\bar{y} \in \Omega \mid \text{for every } y \in Y \text{ there is a } \bar{\lambda} > 0 \text{ with}$$
$$\bar{y} + \lambda y \in \Omega \text{ for all } \lambda \in [0, \bar{\lambda}]\} .$$

The set of all elements in Y which do not belong to $\text{cor}(\Omega)$ and $\text{cor}(Y \setminus \Omega)$ is called the *algebraic boundary* of Ω. If Ω is a nonempty convex set of a real topological linear space with nonempty topological interior $\text{int}(\Omega)$, then, by Lemma 1.32 in [94],

$$\text{int}(\Omega) = \text{cor}(\Omega).$$

Lemma 1.7 ([94, Lemmas 1.12, 1.26 and 3.21]). *Let K be a convex cone in a real linear space Y.*

(i) It holds

$$cor(K) \subset \{y \in Y \mid y'(y) > 0 \text{ for all } y' \in K' \setminus \{0_{Y'}\}.$$

(ii) If Y is locally convex and K is closed, then

$$K = \{y \in Y \mid y^*(y) \geq 0 \text{ for all } y^* \in K^*\}.$$

(iii) If Y is a real topological linear space and $int(K) \neq \emptyset$, then

$$int(K) = \{y \in Y \mid y^*(y) > 0 \text{ for all } y^* \in K^* \setminus \{0_{Y^*}\}\}.$$

(iv) If $cor(K) \neq \emptyset$ then

$$cor(K) = K + cor(K).$$

In classical vector optimization, the optimality concepts are based on partial orderings. More general concepts allow to compare elements $z \in Y$ with some element $y \in Y$ with respect to (w.r.t.) the ordering cone defined to the element y, or, alternatively, w.r.t. the ordering cone defined to each element z which should be compared. These more general concepts are the topic of this book. For both concepts we need to define a convex cone $\mathcal{D}(y) \subset Y$ (and thus a partial ordering) to each element $y \in Y$. Therefore, in the following, we assume a set-valued map $\mathcal{D}: Y \to 2^Y$ to be given with $\mathcal{D}(y)$ a nonempty convex cone for all $y \in Y$. This leads to the following two relations $\leq_1$ and $\leq_2$ and to the definition of a variable ordering structure. Let $y, z \in Y$. We define

$$y \leq_1 z \quad :\Leftrightarrow \quad z - y \in \mathcal{D}(y) \tag{1.1}$$

as well as

$$y \leq_2 z \quad :\Leftrightarrow \quad z - y \in \mathcal{D}(z). \tag{1.2}$$

Definition 1.8. Let Y be a real linear space and $\mathcal{D}: Y \to 2^Y$ a set-valued map with $\mathcal{D}(y)$ a nonempty convex cone for all $y \in Y$. If elements in the space Y are compared using the binary relation (1.1) or (1.2), then the cone-valued map $\mathcal{D}$ is called an *ordering map* and it is said that $\mathcal{D}$ defines a *variable ordering (structure)* on Y.

Example 1.9. Let Y be the Euclidean space $\mathbb{R}^2$ equipped with a variable ordering structure defined by the ordering map $\mathcal{D}: \mathbb{R}^2 \to 2^{\mathbb{R}^2}$ with $\mathcal{D}(y) = \mathbb{R}^2_+$ for all $y \in \mathbb{R}^2 \setminus \{(0,0)\}$ and $\mathcal{D}(0,0) = \text{cone conv}\{(1,1),(1,0)\}$.

Then for all $y, z \in \mathbb{R}^2$ with $y \neq 0_{\mathbb{R}^2}$

$$y \leq_1 z \quad \Leftrightarrow \quad y_i \leq z_i \text{ for } i = 1, 2.$$

For $y = 0_{\mathbb{R}^2}$ and arbitrary $z \in \mathbb{R}^2$ we obtain

$$y = 0_{\mathbb{R}^2} \leq_1 z \quad \Leftrightarrow \quad z \in \text{cone conv}\{(1, 1), (1, 0)\}.$$

For the second relation, relation $\leq_2$, we get for all $y, z \in \mathbb{R}^2$ with $z \neq 0_{\mathbb{R}^2}$

$$y \leq_2 z \quad \Leftrightarrow \quad y_i \leq z_i \text{ for } i = 1, 2.$$

For $z = 0_{\mathbb{R}^2}$ and arbitrary $y \in \mathbb{R}^2$ we obtain

$$y \leq_2 z = 0_{\mathbb{R}^2} \quad \Leftrightarrow \quad -y \in \text{cone conv}\{(1, 1), (1, 0)\}.$$

Of course, if $\mathcal{D}(y) = K$ for all $y \in Y$ for some convex cone K both binary relations (1.1) and (1.2) define the same partial ordering on Y. However, in general the relations do not define a partial ordering:

Lemma 1.10. *Let Y be a real linear space with an ordering map $\mathcal{D}$.*

(i) *The relations defined in (1.1) and (1.2) are reflexive.*
(ii) *The binary relation $\leq_1$ defined in (1.1) is transitive if*

$$\mathcal{D}(y + d) \subset \mathcal{D}(y) \text{ for all } y \in Y \text{ and for all } d \in \mathcal{D}(y). \qquad (1.3)$$

If $\mathcal{D}(y)$ is algebraically closed for all $y \in Y$, then (1.3) also is necessary for the transitivity of $\leq_1$.
(iii) *The binary relation $\leq_2$ defined in (1.2) is transitive if*

$$\mathcal{D}(y - d) \subset \mathcal{D}(y) \text{ for all } y \in Y \text{ and for all } d \in \mathcal{D}(y). \qquad (1.4)$$

If $\mathcal{D}(y)$ is algebraically closed for all $y \in Y$, then (1.4) also is necessary for the transitivity of $\leq_2$.
(iv) *The property given in Definition 1.1(b)(iii) (compatibility with addition) is satisfied by any of the two relations $\leq_1$ or $\leq_2$ if and only if $\mathcal{D}$ is a constant map.*
(v) *The property given in Definition 1.1(b)(iv) (compatibility with nonnegative scalar multiplication) is satisfied by any of the two relations $\leq_1$ or $\leq_2$ if and only if*

$$\mathcal{D}(y) \subset \mathcal{D}(\alpha\, y) \text{ for all } y \in Y \text{ and for all } \alpha > 0. \qquad (1.5)$$

(vi) *The relations defined in (1.1) and (1.2) are antisymmetric if $\mathcal{D}(Y) := \bigcup_{y \in Y} \mathcal{D}(y)$ is pointed.*

Proof.

(i) The relations are both reflexive as the sets $\mathcal{D}(y)$ are assumed to be cones and thus $0_Y \in \mathcal{D}(y)$ for all $y \in Y$.

(ii) We first show that the condition (1.3) is sufficient. Let $x, y, z \in Y$ be arbitrarily given. As $x \leq_1 y$ and $y \leq_1 z$ correspond to $y - x \in \mathcal{D}(x)$ and $z - y \in \mathcal{D}(y)$, (1.3) implies $\mathcal{D}(y) \subset \mathcal{D}(x)$ and we get $z - x = (z - y) + (y - x) \in \mathcal{D}(y) + \mathcal{D}(x) \subset \mathcal{D}(x)$ and hence $x \leq_1 z$.

Next, we show that condition (1.3) also is necessary if $\mathcal{D}(y)$ is algebraically closed for all $y \in Y$. For that we assume $\leq_1$ to be transitive, but (1.3) does not hold. Then there exists some $x \in Y$ and some $d \in \mathcal{D}(x)$ as well as some

$$k \in \mathcal{D}(x + d) \setminus \{0_Y\} \text{ with } k \notin \mathcal{D}(x). \tag{1.6}$$

For all $s > 0$ we obtain $sk \in \mathcal{D}(x + d) \setminus \{0_Y\}$ and $sk \notin \mathcal{D}(x)$. We set

$$y := x + d \text{ and } z_s := y + sk = x + d + sk \text{ for all } s > 0.$$

Then $y - x = d \in \mathcal{D}(x)$ and $z_s - y = sk \in \mathcal{D}(x + d) = \mathcal{D}(y)$ for all $s > 0$. Because $\leq_1$ is transitive, it holds $z_s - x = d + sk \in \mathcal{D}(x)$ for all $s > 0$, i.e. $\frac{1}{s}d + k \in \mathcal{D}(x)$ for $s > 0$ implying, because $\mathcal{D}(x)$ is algebraically closed, $k \in \mathcal{D}(x)$ in contradiction to (1.6).

(iii) We first show that the condition (1.4) is sufficient. Let $x, y, z \in Y$ be arbitrarily given. As $x \leq_2 y$ and $y \leq_2 z$ correspond to $y - x \in \mathcal{D}(y)$ and $z - y \in \mathcal{D}(z)$, (1.4) implies $\mathcal{D}(y) \subset \mathcal{D}(z)$ and we get $z - x = (z - y) + (y - x) \in \mathcal{D}(z) + \mathcal{D}(y) \subset \mathcal{D}(z)$ and hence $x \leq_1 z$.

Next, we show that condition (1.4) also is necessary if $\mathcal{D}(y)$ is algebraically closed for all $y \in Y$. For that we assume $\leq_2$ is transitive, but (1.4) does not hold. Then there exists some $z \in Y$ and some $d \in \mathcal{D}(z)$ as well as some $k \in Y \setminus \{0_Y\}$ with

$$sk \in \mathcal{D}(z - d) \setminus \{0_Y\} \text{ and } sk \notin \mathcal{D}(z) \text{ for all } s > 0. \tag{1.7}$$

We set $y := z - d$ and $x_s := y - sk = z - d - sk$ for $s > 0$ and obtain $y - x_s = sk \in \mathcal{D}(z - d) = \mathcal{D}(y)$ and $z - y = d \in \mathcal{D}(z)$ for all $s > 0$. Because $\leq_2$ is transitive, it holds $z - x_s = d + sk \in \mathcal{D}(z)$ for all $s > 0$ implying $k \in \mathcal{D}(z)$ in contradiction to (1.7).

(iv) The property given in Definition 1.1(b)(iii) corresponds for both relations to the property $\mathcal{D}(y) + \mathcal{D}(z) \subset \mathcal{D}(y + z)$ for any $y, z \in Y$, i.e. to the subadditivity of the cone-valued map $\mathcal{D}$. Using Lemma 3.23 of Chap. 3 the conclusion follows.

(v) As $\mathcal{D}(y)$ is a cone for all $y \in Y$ it holds $\mathcal{D}(y) = \alpha \mathcal{D}(y)$ for all $\alpha > 0$ and thus the property given in Definition 1.1(b)(iii) corresponds for both relations to the property $\mathcal{D}(y) \subset \mathcal{D}(\alpha y)$ for all $y \in Y$ and all $\alpha > 0$.

(vi) $y \leq_1 z$ and $z \leq_1 y$ are equivalent to $z \in \{y\} + \mathcal{D}(y)$ and $z \in \{y\} - \mathcal{D}(z)$, thus $z - y \in \mathcal{D}(Y) \cap (-\mathcal{D}(Y))$, i.e. $z = y$. Analogously for $\leq_2$. $\square$

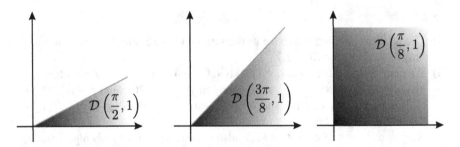

Fig. 1.1 Some of the cones $\mathcal{D}(y)$ of Example 1.11

We speak of a variable ordering (structure) given by the ordering map $\mathcal{D}$, even though the binary relations $\leq_1$, $\leq_2$ given above are in general neither transitive nor compatible with the linear structure of the space. By that we emphasize that here the partial ordering defined by a convex cone K is replaced by a relation which is defined by a cone-valued map $\mathcal{D}$. Note that for instance condition (1.3) can also be written as $\mathcal{D}(y + d) + \mathcal{D}(y) \subset \mathcal{D}(y)$ for all $y \in Y$ and all $d \in \mathcal{D}(y)$ as $\mathcal{D}(y)$ are convex cones for all $y \in Y$. A variable ordering structure satisfying this condition is given in the following example.

Example 1.11. Let Y be the Euclidean space $\mathbb{R}^2$ equipped with a variable ordering structure defined by the ordering map $\mathcal{D}: \mathbb{R}^2 \to 2^{\mathbb{R}^2}$ with

$$
\mathcal{D}(y_1, y_2) := \begin{cases} \left\{ \begin{pmatrix} r \cos \varphi \\ r \sin \varphi \end{pmatrix} \in \mathbb{R}^2 \ \middle| \ r \geq 0, \ \varphi \in [0, \frac{\pi}{8}] \right\} & \text{if } y_1 \geq \frac{\pi}{2}, \\[4mm] \left\{ \begin{pmatrix} r \cos \varphi \\ r \sin \varphi \end{pmatrix} \in \mathbb{R}^2 \ \middle| \ r \geq 0, \ \varphi \in [0, \frac{\pi}{2} + \frac{\pi}{8} - y_1] \right\} & \\[2mm] & \hspace{-2cm} \text{if } y_1 \in (\frac{\pi}{8}, \frac{\pi}{2}), \\[3mm] \mathbb{R}^2_+ & \text{if } y_1 \leq \frac{\pi}{8}, \end{cases}
$$

see Fig. 1.1.

It holds $\mathcal{D}(y) \subset \mathbb{R}^2_+$ and $\mathcal{D}(y)$ is a closed pointed convex cone for all $y \in \mathbb{R}^2$. $\mathcal{D}(y)$ depends on y_1 only and for $z_1 \geq y_1$ for some $y, z \in \mathbb{R}^2$ it holds $\mathcal{D}(z) \subset \mathcal{D}(y)$. As for any $y \in \mathbb{R}^2$ and any $d \in \mathcal{D}(y)$ we have $d_1 \geq 0$ and thus $y_1 + d_1 \geq y_1$ we conclude that (1.3) is satisfied.

A variable ordering structure satisfying instead condition (1.4) is provided in the following example.

Example 1.12. Let Y be the Euclidean space $\mathbb{R}^2$ equipped with a variable ordering structure defined by the ordering map $\mathcal{D}: \mathbb{R}^2 \to 2^{\mathbb{R}^2}$ with

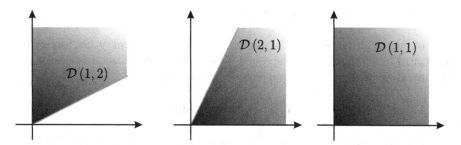

Fig. 1.2 Some of the cones $\mathcal{D}(y)$ of Example 1.13, cf. [48]

$$\mathcal{D}(y_1, y_2) := \begin{cases} \left\{ \begin{pmatrix} r\cos\varphi \\ r\sin\varphi \end{pmatrix} \in \mathbb{R}^2 \,\middle|\, r \geq 0, \; \varphi \in [0, \tfrac{\pi}{8}] \right\} & \text{if } y_1 \leq \tfrac{\pi}{8}, \\[2mm] \left\{ \begin{pmatrix} r\cos\varphi \\ r\sin\varphi \end{pmatrix} \in \mathbb{R}^2 \,\middle|\, r \geq 0, \; \varphi \in [0, y_1] \right\} & \text{if } y_1 \in (\tfrac{\pi}{8}, \tfrac{\pi}{2}), \\[2mm] \mathbb{R}^2_+ & \text{if } y_1 \geq \tfrac{\pi}{2}. \end{cases}$$

Again, $\mathcal{D}(y) \subset \mathbb{R}^2_+$ and $\mathcal{D}(y)$ is a closed pointed convex cone for all $y \in \mathbb{R}^2$. $\mathcal{D}(y)$ depends on y_1 only and for $z_1 \leq y_1$ for some $y, z \in \mathbb{R}^2$ it holds $\mathcal{D}(z) \subset \mathcal{D}(y)$. As for any $y \in \mathbb{R}^2$ and any $d \in \mathcal{D}(y)$ we have $d_1 \geq 0$ and thus $y_1 - d_1 \leq y_1$ we conclude that (1.4) is satisfied. Also, as $y_1 \leq y_1 + d_1$, we obtain that

$$\mathcal{D}(y) \subset \mathcal{D}(y + d) \quad \text{for all } y \in Y \text{ and for all } d \in \mathcal{D}(y). \tag{1.8}$$

In the following we provide an example for an ordering map $\mathcal{D}$ satisfying the condition (1.5) and being thus compatible with nonnegative scalar multiplication.

Example 1.13. Let Y be the Euclidean space $\mathbb{R}^2$ equipped with a variable ordering structure defined by the ordering map $\mathcal{D}: \mathbb{R}^2 \to 2^{\mathbb{R}^2}$ with

$$\mathcal{D}(y) = \left\{ \begin{pmatrix} r\cos\varphi \\ r\sin\varphi \end{pmatrix} \in \mathbb{R}^2 \,\middle|\, r \geq 0, \; \varphi \in \left[\bar{\varphi}_y - \frac{\pi}{4}, \bar{\varphi}_y + \frac{\pi}{4}\right] \cap \left[0, \frac{\pi}{2}\right] \right\}$$

if $y \neq 0_{\mathbb{R}^2}$ and $\mathcal{D}(0_{\mathbb{R}^2}) = \mathbb{R}^2_+$, with $\bar{\varphi}_y \in [0, \pi/2)$ defined by

$$y = (r_y \cos(l\bar{\varphi}_y), r_y \sin(l\bar{\varphi}_y)) \text{ for some } l \in \mathbb{N} \text{ and some } r_y \in \mathbb{R}, \; r_y > 0.$$

For an illustration of some of these cones see Fig. 1.2.

Then $\mathcal{D}(y) = \mathcal{D}(y/\|y\|_2)$ for all $y \in \mathbb{R}^2 \setminus \{0_{\mathbb{R}^2}\}$ and thus $\mathcal{D}(y) = \mathcal{D}(\alpha y)$ for all $\alpha > 0$ and all $y \in Y$.

The binary relation defined in (1.1) describes the idea of domination: all elements of the set

$$\{y\} + \mathcal{D}(y) \setminus \{0_Y\} = \{z \in Y \setminus \{y\} \mid y \leq_1 z\}$$

are considered to be worse (less preferred) than the element y and are thus dominated by y. Here, one is interested in those elements which are worse than the current element y. Then we can interpret the cones $\mathcal{D}(y)$ as

$$\mathcal{D}(y) = \{d \in Y \mid y + d \text{ is worse than } y\} \cup \{0_Y\}.$$

On the other hand, the relation (1.2) corresponds to the concept of preference, as all elements of the set

$$\{y\} - \mathcal{D}(y) \setminus \{0_Y\} = \{z \in Y \setminus \{y\} \mid z \leq_2 y\}$$

are considered to be better or more preferred than the element y. Here, one looks on which elements are better than the current element y. Therefore, it is more natural to define in a first step a cone-valued map $\mathcal{P}: Y \to 2^Y$ with $\mathcal{P}(y)$ a convex cone for all $y \in Y$ and

$$\mathcal{P}(y) = \{d \in Y \mid y + d \text{ is preferred to } y\} \cup \{0_Y\}.$$

Then

$$y \leq z \quad \Leftrightarrow \quad y \in \{z\} + \mathcal{P}(z)$$

for all $y, z \in Y$. By defining $\mathcal{D}(y) := -\mathcal{P}(y)$ for all $y \in Y$ we get a unified notation and the binary relation $\leq_2$ by

$$y \leq_2 z \quad \Leftrightarrow \quad y \in \{z\} - \mathcal{D}(z)$$

for all $y, z \in Y$.

We use here a unified notation for which it holds in case of $\mathcal{D}(y) = K$ for all $y \in Y$ and K a convex cone that $\leq_1$ and $\leq_2$ coincide and are equivalent to a partial ordering introduced by the ordering cone K. However, note that the underlying concepts are fundamentally different and that in general

$$\{d \in Y \mid y + d \text{ is worse than } y\} \neq -\{d \in Y \mid y + d \text{ is preferred to } y\}.$$

1.2   Variable Ordering Structures Given by Bishop-Phelps Cones

Variable ordering structures on a real linear space are defined by a set-valued map with images being convex cones. In classical vector optimization with a partial ordering one often assumes the partial ordering to be additional antisymmetric and thus the related ordering cone has to be pointed. Similarly, we may assume the cones $\mathcal{D}(y)$ to be pointed for all $y \in Y$. Also, for numerical reasons and as it is in general the case in applications, we may further assume the cones $\mathcal{D}(y)$ to be closed. In finite dimensions all closed pointed convex cones are representable as so-called Bishop-Phelps cones, see Theorem 1.17 below.

In the following we give the definition of such cones and we focus on variable ordering structures in a real normed space given by Bishop-Phelps cones, i.e. cone-valued maps with images being Bishop-Phelps cones. The advantage of doing this will become obvious in Chap. 6. There, we examine scalarization functionals for vector optimization problems with such a variable ordering structure which uses the special structure of the Bishop-Phelps cones.

1.2.1 Bishop-Phelps Cones

As the definition of Bishop-Phelps cones requires a norm, we assume Y to be a real normed space with the dual space denoted by Y^*. Here, $\| \cdot \|$ and $\| \cdot \|_*$ denote the norms in Y and Y^*, respectively, with

$$\|\phi\|_* := \sup_{y \neq 0_Y} \frac{|\phi(y)|}{\|y\|} \text{ for all } \phi \in Y^*.$$

A Bishop-Phelps cone is defined by an element ϕ from the dual space Y^* as follows:

Definition 1.14. For an arbitrary continuous linear functional ϕ on the normed space Y the cone

$$C(\phi) := \{y \in Y \mid \|y\| \leq \phi(y)\} \tag{1.9}$$

is called *Bishop-Phelps cone* (BP cone).

A cone $K \subset Y$ for which a functional $\phi \in Y^*$ and a norm $\| \cdot \|$ equivalent to the norm of the space exist such that K can be written as in (1.9) is called *representable as a BP cone*.

Note that the original concept of a BP cone is slightly different from the above one. Namely, for an arbitrary continuous linear functional ϑ on the normed space Y with $\|\vartheta\|_* = 1$ the cone

$$\{y \in Y \mid t \, \|y\| \leq \vartheta(y)\} \tag{1.10}$$

with some scalar $t \in (0, 1)$ was originally considered. It is easy to see that any cone satisfying (1.10) is also a BP cone in the sense of Definition 1.14 with $\phi = \vartheta/t$, and any BP cone in the sense of Definition 1.14 with $\|\phi\|_* > 1$ satisfies (1.10) with $\vartheta = \phi/\|\phi\|_*$ and $t = 1/\|\phi\|_*$. We use in the following BP cones in the sense of Definition 1.14.

Fig. 1.3 BP cone $C(\phi_1, \phi_2)$ of Example 1.15(c) for $\phi_1 = 2$ and $\phi_2 = 3/2$, as well as the unit ball w.r.t. the l_1-norm and (in *dashed line*) the set $\{(y_1, y_2) \in \mathbb{R}^2 \mid (\phi_1, \phi_2)^\top (y_1, y_2) = 1\}$, cf. [52]

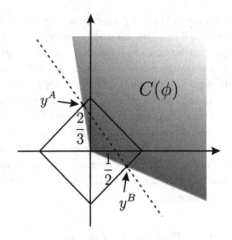

Example 1.15. (a) [93] Let $Y = \mathbb{R}^n$ and $\| \cdot \|_p$ be an l_p norm with $p \in [1, \infty]$. The cone

$$C_p := \{y \in \mathbb{R}^n \mid \|(y_1, \ldots, y_{n-1})\|_p \le y_n\}$$

is representable as a BP cone by $C_p = C(\sqrt[p]{2}e_n)$ for $p \in [1, \infty)$ and $e_n := (0, \ldots, 0, 1)^\top$, and $C_\infty = C(e_n)$. The cone C_2 is the Lorentz cone, also called second-order cone or ice cream cone, which is a well-known concept in second-order cone programming.

(b) [93] The natural ordering cones in the Banach spaces $L_1([0, 1])$ and l^1 are representable as BP cones.

(c) Let $Y = \mathbb{R}^2$ and assume that the space is equipped with the l_1-norm. Then for instance for $(\phi_1, \phi_2) = (1, 1)$ we have $C(\phi_1, \phi_2) = \mathbb{R}_+^2$. Assume $\phi_1, \phi_2 \ge 1$, then $\mathbb{R}_+^2 \subset C(\phi_1, \phi_2)$, $(0, 1/\phi_2) \in C(\phi_1, \phi_2)$, $(1/\phi_1, 0) \in C(\phi_1, \phi_2)$ and

$$C(\phi_1, \phi_2) = \text{cone conv}\{y^A, y^B\}$$

with

$$y^A := \left(\frac{1 - \phi_2}{\phi_1 + \phi_2}, \frac{1 + \phi_1}{\phi_1 + \phi_2} \right)^\top \quad \text{and} \quad y^B := \left(\frac{1 + \phi_2}{\phi_1 + \phi_2}, \frac{1 - \phi_1}{\phi_1 + \phi_2} \right)^\top,$$

see Fig. 1.3.

Below we collect some properties of BP cones.

Lemma 1.16 ([79, 93]). *Let $\phi \in Y^*$ be given.*

(i) *$C(\phi)$ is closed, pointed and convex.*

(ii) *If $\|\phi\|_* > 1$ then $C(\phi)$ is nontrivial; if $\|\phi\|_* < 1$ then $C(\phi) = \{0_Y\}$.*

(iii) *If $\|\phi\|_* = 1$ then $C(\phi) = \{y \in Y \mid \|y\| = \phi(y)\}$; if, additionally, Y is a reflexive Banach space then $C(\phi)$ is nontrivial.*

(iv) *$\{y \in Y \mid \|y\| < \phi(y)\} \subset int(C(\phi))$.*
If $\|\phi\|_* > 1$ then the interior of $C(\phi)$ is nonempty and

$$int(C(\phi)) = \{y \in Y \mid \|y\| < \phi(y)\}.$$

(v) *$\phi \in C(\phi)^\#$.*

(vi) *The set $\{y \in C(\phi) \mid \phi(y) = 1\}$ is, if nonempty, a closed and bounded base for the cone $C(\phi)$.*

(vii) *$C(\phi)^* = cl(cone(B(\phi, 1)))$ with $B(\phi, 1) = \{y^* \in Y^* \mid \|y^* - \phi\|_* \leq 1\}$.*

The following theorem gives conditions for the representability of a cone as a BP cone.

Theorem 1.17 ([129, Theorems 3.2 and 3.4]). *A nontrivial cone $K \subset Y$ is representable as a BP cone if and only if K is a convex cone with a closed and bounded base. In the Euclidean space $Y = \mathbb{R}^n$ a convex cone $K \subset Y$ is representable as a BP cone if and only if K is closed and pointed.*

Note that there exist important classes of cones which do not have bounded bases. For instance, the natural ordering cones in the spaces l^p and L^p for $1 < p < \infty$ have not a bounded base, see [12, 36].

1.2.2 Augmented Dual Cones

Bishop-Phelps cones are related to augmented dual cones which extend the usual definition of a dual cone. The definition of an augmented dual cone as well as variants as the weak augmented dual cone or the augmented quasi-interior dual cone are given in the following definition together with an illustrative example. As before, let $(Y, \|\cdot\|)$ be a real normed space.

Definition 1.18 ([100]). Let $K \subset Y$ be a closed pointed convex cone.

(a) The set

$$K^{a*} := \{(\phi, \alpha) \in K^\# \times \mathbb{R}_+ \mid \phi(y) - \alpha\|y\| \geq 0 \text{ for all } y \in K\}$$

is called *augmented dual cone*.

(b) Let $int(K) \neq \emptyset$. The set

$$K^{ao} := \{(\phi, \alpha) \in K^\# \times \mathbb{R}_+ \mid \phi(y) - \alpha\|y\| > 0 \text{ for all } y \in int(K)\}$$

is called *weak augmented dual cone*.

(c) The set

$$K^{a\#} := \{(\phi, \alpha) \in K^\# \times \mathbb{R}_+ \mid \phi(y) - \alpha \|y\| > 0 \text{ for all } y \in K \setminus \{0_Y\}\}$$

is called *augmented quasi-interior of the dual cone*.

It holds $K^{a\#} \subset K^{a\circ} \subset K^{a*}$ for any closed pointed convex cone K.

Example 1.19. (a) [100, Example 4.7] Let Y be the Euclidean space $\mathbb{R}^n$ and $K = \mathbb{R}^n_+$. Then

$$K^{a*} = \{(\phi, \alpha) \in \text{int}(\mathbb{R}^n_+) \times \mathbb{R}_+ \mid \phi_i \geq \alpha, \ i = 1, \ldots, n\}.$$

To see this, note that $(\phi, \alpha) \in K^{a*}$ if and only if $\alpha \geq 0$, $\phi_i > 0$, $i = 1, \ldots, n$ and

$$\phi^\top y - \alpha \|y\|_2 \geq 0 \text{ for all } y \in \mathbb{R}^n_+.$$

If we evaluate this inequality for vectors $y \in \mathbb{R}^n_+$ with $y_i = 0$ for all $i \in \{1, \ldots, n\} \setminus \{j\}$ and $y_j > 0$, we obtain the necessary condition $\phi_j \geq \alpha$ for all $j \in \{1, \ldots, n\}$. For $(\phi, \alpha) \in \text{int}(\mathbb{R}^n_+) \times \mathbb{R}_+$, this necessary condition is also sufficient for $(\phi, \alpha) \in K^{a*}$, as it implies for any $y \in \mathbb{R}^n_+$

$$\phi^\top y - \alpha \|y\|_2 \geq \alpha \|y\|_1 - \alpha \|y\|_2 \geq 0.$$

(b) Let Y be the Euclidean space $\mathbb{R}^n$ and let $K = \{y \in \mathbb{R}^n \mid y = Bx, \ x \in \mathbb{R}^m_+\}$ be a polyhedral cone with $B \in \mathbb{R}^{n \times m}$ a matrix. Then for the set

$$\tilde{K} := \{(\phi, \alpha) \in \mathbb{R}^n \times \mathbb{R}_+ \mid (B^\top \phi)_i > 0, \\ (B^\top \phi)_i \geq \alpha \|B\|_*, \ i = 1, \ldots, m\} \tag{1.11}$$

it holds $\tilde{K} \subset K^{a*}$. This holds because $(\phi, \alpha) \in K^{a*}$ if and only if $\alpha \geq 0$, $\phi \in K^\#$ and

$$(B^\top \phi)^\top x - \alpha \|Bx\|_2 \geq 0 \text{ for all } x \in \mathbb{R}^m_+.$$

A sufficient condition for that is $B^\top \phi \in (\mathbb{R}^n_+)^\#$ and $(B^\top \phi, \alpha \|B\|_*) \in (\mathbb{R}^n_+)^{a*}$. With part (a) of this example we obtain (1.11).

For the definition of the augmented dual cones, it is crucial that the quasi-interior of the dual cone $K^\#$ is nonempty. For instance the Krein-Rutman theorem, below as cited in [94, Theorem 3.38], gives conditions ensuring that:

Theorem 1.20 (Krein-Rutman Theorem). *In a real separable normed space $(Y, \|\cdot\|)$ with a closed and pointed convex cone $K \subset Y$ the quasi interior $K^\# = K^\#_{Y*}$ of the topological dual cone is nonempty.*

Thus, in the finite dimensional Euclidean space $Y = \mathbb{R}^n$ for any closed pointed convex cone the quasi interior of the dual cone is nonempty and according to [158], see also [82, p. 199], it equals the interior of the dual cone. The pointedness of K is essential as for any convex cone K the condition $K^{\#} \neq \emptyset$ already implies the pointedness of K [94, Lemma 1.27]. There is also a strong relation between the base of a cone and the elements of the quasi-interior of the dual cone. A convex cone has a base if and only if the quasi-interior of the dual cone is nonempty and in this case for any element $y^* \in K^{\#}$ the set

$$\{y \in K \mid y^*(y) = 1\}$$

is a base of K [94, Lemma 1.28]. For BP cones $C(\phi)$ the quasi-interior of the dual cone is nonempty according to Lemma 1.16(v). We obtain the following relation of BP cones and elements of the augmented dual cones:

Lemma 1.21. *Let $\phi \in Y^*$ be given which defines a BP cone*

$$C(\phi) = \{y \in Y \mid \|y\| \leq \phi(y)\}.$$

Then

$$(\phi, \alpha) \in (C(\phi))^{a*} \text{ for all } \alpha \in [0, 1]$$

and $(\phi, 0) \in (C(\phi))^{a\#}$.
If $\|\phi\|_* > 1$ then

$$(\phi, \alpha) \in (C(\phi))^{a\circ} \text{ for all } \alpha \in [0, 1].$$

Proof. With Lemma 1.16(v) $\phi \in C(\phi)^{\#}$. According to the definition of the BP cone, for $\alpha \in [0, 1]$ it holds

$$0 \leq \phi(y) - \|y\| \leq \phi(y) - \alpha\|y\| \text{ for all } y \in C(\phi)$$

and thus $(\phi, \alpha) \in C(\phi)^{a*}$. Also, as $\phi(y) > 0$ for all $y \in C(\phi) \setminus \{0_Y\}$, $(\phi, 0) \in (C(\phi))^{a\#}$. According to Lemma 1.16(iv), if $\|\phi\|_* > 1$ and $\alpha \in [0, 1]$,

$$0 < \phi(y) - \|y\| \leq \phi(y) - \alpha\|y\| \text{ for all } y \in \text{int}(C(\phi))$$

which implies $(\phi, \alpha) \in (C(\phi))^{a\circ}$. $\square$

Lemma 1.22. *Let $K \subset Y$ be a nontrivial closed convex pointed cone and $C(\phi)$ a BP cone. If $K \subset C(\phi)$, then $(\phi, 1) \in K^{a*}$ and $(\phi, 0) \in K^{a\#}$. If additionally $\text{int}(K) \neq \emptyset$ and $\|\phi\|_* > 1$, then $(\phi, 1) \in K^{a\circ}$.*

Proof. From the definitions it follows $(C(\phi))^{aN} \subset K^{aN}$ for any $N \in \{*, \circ, \#\}$. The conclusion is then an application of Lemma 1.21.   □

Cones $K \subset Y$ having the property assumed in the above lemma are called supernormal:

Definition 1.23 ([71]). Let $K \subset Y$ be a nontrivial convex cone. K is said to be *supernormal* (or nuclear or has the angle property) if there exists $\phi \in Y^*$ such that

$$K \subset \{y \in Y \mid \|y\| \leq \phi(y)\} = C(\phi). \tag{1.12}$$

For instance, in $\mathbb{R}^n$ every pointed convex cone is supernormal and every BP cone is a supernormal cone [91, p. 635]. For closed pointed supernormal cones we immediately obtain by Lemma 1.22 that the augmented dual cones are nonempty:

Corollary 1.24. *Let $K \subset Y$ be a nontrivial closed pointed convex supernormal cone. Then there exists $\phi \in Y^*$ such that (1.12) holds and $(\phi, 1) \in K^{a*}$, $(\phi, 0) \in K^{a\#}$. If additionally $int(K) \neq \emptyset$ and $\|\phi\|_* > 1$, then $(\phi, 1) \in K^{a\circ}$.*

This leads to the following result.

Lemma 1.25. *Let $K \subset Y$ be a nontrivial closed pointed convex cone. Then the following is equivalent:*

(i) There exists $(\phi, \alpha) \in K^{a*}$ with $\alpha \neq 0$.
(ii) K is supernormal.

Proof. (ii) implies (i) by Corollary 1.24. (i) implies that $(\tilde{\phi}, 1) \in K^{a*}$ with $\tilde{\phi} := \frac{1}{\alpha}\phi$, i.e. $\tilde{\phi}(y) - \|y\| \geq 0$ for all $y \in K$ and hence $K \subset C(\tilde{\phi})$ and K is supernormal.
                                                                                           □

In a real normed space a cone is supernormal if and only if it is representable as a BP cone:

Lemma 1.26 ([91]). *Let K be a nontrivial closed pointed convex cone. Then the following is equivalent:*

(i) K is representable as a BP cone.
(ii) K is supernormal.

Proof. According to [91, p. 635] any BP cone is a supernormal cone (denoted nuclear in [91]). On the other hand, (ii) implies for a convex cone that the cone has a bounded base, see [91, p. 635] and the references therein. With Theorem 1.17, see also Theorem 1 in [91], K is representable as a BP cone.   □

Hence, the augmented dual cone of some cone contains elements (ϕ, α) with $\alpha \neq 0$ if and only if the cone is representable as a BP cone.

1.2.3 Variable Ordering Structures Given by Bishop-Phelps Cones

In this section we assume that the variable ordering structure on the real normed space Y is defined by a set-valued map $\mathcal{D}: Y \rightarrow 2^Y$ with each cone $\mathcal{D}(y)$ representable as a BP cone. Then, with any $y \in Y$ we associate a norm $\| \cdot \|_y$ equivalent to but eventually different from the norm of the space Y and we define a map $\ell: Y \rightarrow Y^*$ such that

$$\mathcal{D}(y) = C(\ell(y)) = \{u \in Y \mid \|u\|_y \leq \ell(y)(u)\} \text{ for all } y \in Y.$$

Recall that it is not a strong restriction to assume the images of the ordering map to be representable as BP cones. In most literature, the convex cones appearing in vector optimization are closed and pointed, for numerical reasons or due to the application. According to Theorem 1.17, in finite dimensions such cones are all representable as BP cones.

But note that already in $\mathbb{R}^n$ one might need different equivalent norms to represent different nontrivial closed pointed convex cones as BP cones. In $\mathbb{R}^2$ it satisfies to choose just one norm but already in $\mathbb{R}^3$ one has to use different norms to model for instance a polyhedral cone and the Lorentz cone. In an application however, there might be an ordering map with different cones $\mathcal{D}(y)$ but presumably they will all be of the same type, for instance all polyhedral, and can all be modeled with the same norm.

In [57, 58] properties are studied which should naturally be satisfied by variable ordering structures which describe the preferences of a decision maker in an Euclidean space. A special structure is proposed for the images $\mathcal{D}(y)$ which implies that these cones are BP cones w.r.t. the Euclidean norm, see [57, Remark 3.2.17] and [58, Remark 8]: among others, the cones $\mathcal{D}(y) \subset \mathbb{R}^m$ shall be closed, convex and so-called ideal-symmetric cones with $\mathbb{R}^m_+ \subset \mathcal{D}(y)$ for all $y \in A$ with A a nonempty subset of $\mathbb{R}^m$. Here, ideal symmetric means that the cone $\mathcal{D}(y)$ is some kind of symmetric w.r.t. the direction $y - z$ pointing to an ideal point $z \in \mathbb{R}^m$ which is a point z with $z_i \leq \inf\{y_i \mid y \in A\}$ for $i = 1, \ldots, m$. More details are provided in the next example.

Example 1.27. Consider the cone-valued map $\mathcal{D}: A \rightarrow 2^{\mathbb{R}^m}$ on some bounded set $A \subset \mathbb{R}^m$ defined by

$$\mathcal{D}(y) := \{d \in \mathbb{R}^m \mid d^\top (y - p) \geq \gamma \cdot \|d\|_2 \cdot [y - p]_{\min}\} \text{ for all } y \in A$$

where $\gamma \in (0, 1]$, $p_i < \inf_{y \in A} y_i$ for $i = 1, \ldots, m$, and

$$[y - p]_{\min} := \min_{i=1,\ldots,m} y_i - p_i,$$

compare [58, 86].

Here, $\gamma \in (0, 1]$ is a scalar which controls the angle of the cone and $p \in \mathbb{R}^m$ is chosen as ideal vector. These cones are Bishop-Phelps cones what can be seen by setting

$$\ell(y) := \frac{1}{\gamma \, [y - p]_{\min}} (y - p)$$

and

$$\mathcal{D}(y) = C(\ell(y)) := \{u \in Y \mid \|u\|_2 \leq \ell(y)^\top u\}. \tag{1.13}$$

Moreover, $\mathbb{R}^m_+ \subset \mathcal{D}(y)$ for all $y \in A$, because for any $d \in \mathbb{R}^m_+ \setminus \{0_{\mathbb{R}^m}\}$ it holds

$$\frac{d^\top (y - p)}{\gamma \|d\|_2 [y - p]_{\min}} \geq \frac{d^\top (y - p)}{\gamma \|d\|_1 [y - p]_{\min}} \geq \frac{\|d\|_1 [y - p]_{\min}}{\gamma \|d\|_1 [y - p]_{\min}} = \frac{1}{\gamma} \geq 1,$$

i.e. $d \in \mathcal{D}(y)$.

In particular, when the norm $\| \cdot \|_y$ in the definition of the BP cones $\mathcal{D}(y)$ is assumed to equal the norm $\| \cdot \|$ of the space Y and is thus equal for all $y \in Y$ as in the previous example, these cones reduce to the BP cones

$$\mathcal{D}(y) = C(\ell(y)) = \{u \in Y \mid \|u\| \leq \ell(y)(u)\}. \tag{1.14}$$

Below is an example of an ordering map given by such BP cones. It shows that even if the norm $\| \cdot \|$ does not depend on y a wide range of different cones is covered by the images $\mathcal{D}(y)$ in (1.14).

Example 1.28. Let Y be the Euclidean space $\mathbb{R}^2$, $\| \cdot \|_y := \| \cdot \|_2$ for all $y \in \mathbb{R}^2$ and define $\ell : \mathbb{R}^2 \to \mathbb{R}^2$ by

$$\ell(y_1, y_2) := \left(\frac{3 + \sin y_1}{2}, \frac{3 + \cos y_2}{2} \right)^\top \in [1, 2] \times [1, 2]. \tag{1.15}$$

Then $\mathbb{R}^2_+ \subset C(\ell(y))$ for all $y \in \mathbb{R}^2$. The cones $C(\ell(y))$ can be visualized as follows: The two extreme rays of the pointed convex cone $C(\ell(y))$ are given by two rays starting in the origin being defined by the intersection points of the unit circle and the line connecting the points

$$\left(\frac{1}{\ell_1(y)}, 0 \right) \quad \text{and} \quad \left(0, \frac{1}{\ell_2(y)} \right),$$

see Fig. 1.4. For instance, $C(\ell(3\pi/2, \pi)) = \mathbb{R}^2_+$.

Fig. 1.4 BP cone $C(\ell(y))$ of Example 1.28 for $\ell_1 = \ell_1(y)$ and $\ell_2 = \ell_2(y)$, as well as the unit ball w.r.t. the Euclidean norm and (in *dashed line*) the line connecting the points $(1/\ell_1, 0)$ and $(0, 1/\ell_2)$, cf. [52]

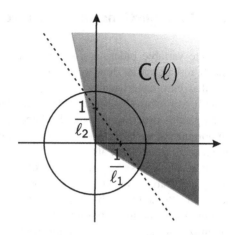

We end this section with an example in the infinite dimensional Hilbert space $L_2([0, 1])$:

Example 1.29. Let $Y = L_2([0, 1])$ denote the real linear space of all (equivalence classes of) quadratic Lebesgue-integrable functions $f : [0, 1] \to \mathbb{R}$ with inner product

$$\langle f, g \rangle := \int_0^1 f(x)g(x)\mathrm{d}x \qquad \forall\ f, g \in L_2([0, 1]).$$

Then Y is a Hilbert space and we can set $Y = Y^*$. Let a map $\ell : Y \to Y^*$ be defined by

$$\ell(f) = f + e \qquad \forall\ f \in L_2([0, 1])$$

with $e \in L_2([0, 1])$, $e(x) := 1$ for all $x \in [0, 1]$. Then a cone-valued map $\mathcal{D} : Y \to 2^Y$ is defined by

$$
\begin{aligned}
\mathcal{D}(f) &:= C(\ell(f)) \\
&= \left\{ g \in L_2([0, 1]) \mid \langle \ell(f), g \rangle \geq \sqrt{\langle g, g \rangle} \right\} \\
&= \left\{ g \in L_2([0, 1]) \ \middle| \ \int_0^1 (f(x)g(x) + g(x))\mathrm{d}x \geq \sqrt{\int_0^1 (g(x))^2 \mathrm{d}x} \right\}
\end{aligned}
$$

for all $f \in L_2([0, 1])$.

1.3 Variable Ordering Structures in Applications

Despite variable ordering structures were already introduced in 1974 [158] not much research was done up to a few years ago on this topic. A main reason for that may have been a lack of published applications which require variable ordering structures for a mathematical modeling. However, already Yu in [159] gave examples of ordering maps and related them to utility functions to illustrate the importance of variable ordering structures in modeling the preferences of decision makers adequately, see also [32].

In recent years the importance of such structures for modeling preferences adequately have been again pointed out. A main contribution in this direction is the above mentioned work by Engau [58] where examples showing the limitations of preference modeling using only a partial ordering are given. Further, in portfolio optimization, and also in location problems, a new notion called equitability is suggested, which is also based on a variable ordering structure. This notion together with a recent application in medical image registration are shortly discussed in this section.

1.3.1 Variable Ordering Structures in Medical Image Registration

In medical image registration it is the aim to match two data sets A and B, derived for instance by X-ray computed tomography (CT) or magnetic resonance tomography (MRT). Hence one searches for a transformation map t among a set of feasible maps T, which should, for some applications, be found automatically without a human decision maker. The quality of the transformation, i.e. the similarity of the transformed data set to the target set, can be measured by several distance measures $f_i: (t, A, B) \rightarrow \mathbb{R}$. Then the aim is to find the transformation with the smallest distance measure.

However, a multitude of measures exist that evaluate distinct characteristics, such as the sum of square differences, mutual information or cross-correlation. Different measures can lead to different best transformations. Some measures fail on special data sets, i.e. they lead to mathematically correct but useless results. Thus it is important to combine several measures. Possible approaches are a weighted sum of different measures. But difficulties appear, such as badly scaled or nonconvex functions.

The approach discussed by Wacker in [146] proposes instead a multiobjective view of this problem by collecting various distance measures in an objective vector $f := (f_1, \ldots, f_m)^\top$ and by solving for given sets A and B a vector optimization problem

$$\min_{t \in T} f(t, A, B).$$

With the Euclidean space $Y = \mathbb{R}^m$ as the objective space of the vector optimization problem, such vector optimization problems are also called *multiobjective optimization* problems. The functions f_i are conflicting scalar-valued functions, also denoted criteria, which all should be minimized at the same time.

It turns out that the natural (componentwise) partial ordering introduced by the nonnegative orthant is not adequate here. Indeed, a transformation t that minimizes one objective function f_i while all the others have relatively high values would be an optimal solution using the componentwise ordering (denoted a weakly efficient solution, see Definitions 2.1 and 2.33) but not a satisfying solution.

To define optimality, a variable ordering structure is thus introduced to incorporate more of the available information in the decision-making. A weighting vector $w = w(y) \in \mathbb{R}_+^m$ is generated for each point $y \in \mathbb{R}^m$ in the objective space. This weight vector depends on gradient information, conformity (equality) and continuity aspects and reflects the preference of a totally rational decision maker who puts a higher weight on promising measures, dependent on the value $f(t, A, B)$ of the actual transformation t. This can be interpreted as some kind of voting between the measures. A weight component equal to zero is also allowed. This corresponds to the omission of a measure, which seems to fail on the data set. A failure means, for instance, that one objective function (measure), which should be minimized, increases for special changes of the current transformation, while all other objective functions decrease, because it evaluates characteristics of the data sets which are misleading.

A weight vector at a point $\bar{y}$ defines a cone of dominated directions by

$$\mathcal{D}^w := \mathcal{D}(w(\bar{y})) := \left\{ d \in \mathbb{R}^m \ \middle| \ \sum_{i=1}^m \mathrm{sgn}(d_i) w_i(\bar{y}) \geq 0 \right\}$$

where

$$\mathrm{sgn}(d_i) := \begin{cases} 1 & \text{if } d_i > 0, \\ 0 & \text{if } d_i = 0, \\ -1 & \text{if } d_i < 0. \end{cases}$$

To illustrate the varying nature of such a variable ordering structure we determine $\mathcal{D}^w \subset \mathbb{R}^2$ for all possible weights $w \in \mathbb{R}_+^2 \setminus \{0_{\mathbb{R}^2}\}$:

$$\mathcal{D}^w := \begin{cases} \{d \in \mathbb{R}^2 \mid d_1 \geq 0, \ d_2 \in \mathbb{R}\} & \text{for } w_1 > 0, \ w_2 = 0, \\ \{d \in \mathbb{R}^2 \mid d_1 \in \mathbb{R}, \ d_2 \geq 0\} & \text{for } w_1 = 0, \ w_2 > 0, \\ \mathbb{R}_+^2 \cup \{d \in \mathbb{R}^2 \mid d_1 < 0 \wedge d_2 > 0\} & \text{for } w_2 \geq w_1 > 0, \\ \mathbb{R}_+^2 \cup \{d \in \mathbb{R}^2 \mid d_1 > 0 \wedge d_2 < 0\} & \text{for } w_1 \geq w_2 > 0. \end{cases}$$

The values w_i of the components of w are also important in higher dimensions ($m \geq 3$). Note that for nonnegative weights $w \in \mathbb{R}_+^m$, the inclusion $\mathbb{R}_+^m \subset \mathcal{D}^w$ holds.

1.3.2 Variable Ordering Structures and Equitability

In portfolio optimization and location problems the concept of equitability is known [103], which is a refinement of the natural (componentwise) partial ordering in finite dimensions. For that concept it is assumed that $Y = \mathbb{R}^m$ is the objective space of a vector optimization problem $\min_{x \in S} f(x)$ for some nonempty set $S \subset \mathbb{R}^n$ and the objective map $f: \mathbb{R}^n \to \mathbb{R}^m$ is given by $f = (f_1, \ldots, f_m)$, with $f_i: \mathbb{R}^n \to \mathbb{R}$, $i = 1, \ldots, m$. For the notion of equitability it is assumed that the criteria f_i are not only uniform in the sense of scale used and directly comparable but also anonymous. This makes the distribution of outcomes $f_i(x)$ for some given $x \in S$ more important than the assignment of several outcomes to the specific criteria f_i.

For instance, in portfolio selection having n securities available, $x_j \in \mathbb{R}_+$ expresses the portion of the capital which is invested in the security j. We consider now m equally probable scenarios. Then c_{ij} denotes the observed (or forecasted) rate of return of security j under scenario i. This results in an outcome matrix $C = (c_{ij})_{ij} \in \mathbb{R}^{m \times n}$. In the portfolio selection problem one considers the linear vector optimization problem

$$\text{Maximize } Cx \text{ subject to } \sum_{j=1}^{n} x_j = 1 \text{ and } x_j \geq 0, \ j = 1, \ldots, n.$$

The objective functions are uniform and it is postulated in [125] that an aggregation must be equitable to model risk averse preferences.

For the concept of equitability, as mentioned above, one is interested in the distribution of the outcomes of the several objectives and not in their ordering, i.e. for instance a vector $(4, 2, 0)$ in the objective space is considered to be equally good as a vector $(0, 4, 2)$. At the same time a principle of transfer should be satisfied stating that a transfer of any small amount from one outcome to any other relatively worse outcome is more preferred. For instance, $(2, 2, 2)$ is considered to be better than $(4, 2, 0)$.

This is modeled in the following way: define the map $\Theta: \mathbb{R}^m \to \mathbb{R}^m$ for any $y \in \mathbb{R}^m$ by

$$\Theta(y) := (\Theta_1(y), \ldots, \Theta_m(y)) \text{ with } \Theta_1(y) \geq \ldots \geq \Theta_m(y)$$

such that there exists a permutation τ of $\{1, \ldots, m\}$ with $\Theta_i(y) = y_{\tau(i)}$ for $i = 1, \ldots, m$. The cumulative map $\overline{\Theta}: \mathbb{R}^m \to \mathbb{R}^m$ is defined by $\overline{\Theta}(y) = (\overline{\Theta}_1(y), \ldots, \overline{\Theta}_m(y))$ with

$$\overline{\Theta}_i(y) = \sum_{j=1}^{i} \Theta_j(y) \text{ for } i = 1, \ldots, m \text{ and for all } y \in \mathbb{R}^m.$$

Then the equitability relation $\preccurlyeq_e$ is defined by

$$x \preccurlyeq_e y \; \Leftrightarrow \; \overline{\Theta}_i(x) \leq \overline{\Theta}_i(y) \; \text{ for all } i = 1, \ldots, m.$$

For instance $\Theta(2, 4, 0) = (4, 2, 0)$ and $\Theta(2, 2, 2) = (2, 2, 2)$ and thus $\overline{\Theta}(2, 4, 0) = (4, 6, 6)$ and $\overline{\Theta}(2, 2, 2) = (2, 4, 6)$. Then $(2, 2, 2) \preccurlyeq_e (2, 4, 0)$. Based on $\preccurlyeq_e$ an *equitable efficient element* $\bar{y}$ of some set $A \subset \mathbb{R}^m$ is defined as an element in A such that there exists no other element $y \in A$ with $y \preccurlyeq_e \bar{y}$. There is a relation to the concept of an efficient element in a partially ordered space ordered by $\mathbb{R}^m_+$, i.e. by the componentwise ordering (for the definition of an efficient element see Definition 2.1): y is an equitable efficient element of the set A if and only if y is an efficient element of the set $\overline{\Theta}(A)$ w.r.t. the ordering cone $\mathbb{R}^m_+$. However, the first $m-1$ objectives of the multiobjective optimization problem $\min_{y \in A} \overline{\Theta}(y)$ are in general non-differentiable, even if the original optimization problem is differentiable, cf. [126, 135].

The problem of finding equitable efficient elements is a vector optimization problem with a variable ordering structure: the space $Y = \mathbb{R}^m$ is partitioned into $m!$ sectors. A sector is a non-pointed convex cone. For each sector a cone or polyhedral set of preferred and of dominated directions is defined, i.e. $\{\mathcal{D}(y) \mid y \in Y\}$ is here a family of a finite number of pointed convex cones or polyhedral sets. The ordering concept of equitability corresponds thus to a variable ordering structure with the sets $\mathcal{D}(y)$ depending on the sector in which the element y is located. For more details we refer to [9].

1.4 Notes on the Literature

Section 1.1 of this chapter contains a collection of standard definitions related to partially ordered linear spaces which can be found in many introductory books on vector optimization as for instance in the books by Göpfert and Nehse [70], Jahn [94] and Luc [113]. Engau gives in [57, Chap. 2] a good survey on the relevant properties of binary relations and the notions of a partial, pre- and total order.

The concept of a variable ordering structure was already introduced by Yu in 1974 [158] in terms of domination structures and further generalized in 1976 with his colleagues Bergstresser and Charnes [19]. Domination structures have soon after been also discussed by Bergstresser and Yu [20] in the context of N person games. Engau [57, 58] calls $\mathcal{D}: Y \rightarrow 2^Y$ with $\mathcal{D}(y)$ a convex cone for all y a variable domination structure. The same in [68] by Giannessi, Mastroeni and Yang. Wiecek [154] speaks of a structure of domination, while Huang et al. [87] use the name variable order relation.

Some examinations of variable ordering structures based on the relation given in (1.2) are done by Chen together with several colleagues in various works starting in 1992 [28–31]. For instance, in the book [29], Chen, Huang and Yang introduce first a very general relation based on the cone-valued map $\mathcal{D}$ by the following: let $w \in Y$, then

$$y \leq_{\mathcal{D}(w)} z \; :\Leftrightarrow \; z - y \in \mathcal{D}(w).$$

For defining optimality they choose $w = z$. We consider here the two special cases $w = y$ and $w = z$, compare (1.1) and (1.2).

In 2007 Huang et al. [87] carried out theoretical examinations regarding variable ordering structures in the context of vector complementarity problems. For a short literature survey about variable ordering structures appearing in the context of vector variational inequalities and vector complementarity problems we refer to Sect. 10.1.

That the importance of criteria may change during the decision-making process and that it may depend on current criteria values was also already recognized by Karaskal and Michalowski in [99]. Wiecek gave an example of this in [154]. Engau lists in [57, p. 48] several authors considering variable ordering structures (or changing preferences) in problems of economic choice and practical decision making. In [102, 132], Korhonen et al. and Ramesh et al. model preferences in multiobjective integer programming by the cones

$$\mathcal{D}(y) = \left\{ \sum_{i=1}^{m} \mu_i (y - y^i) \; \middle| \; \mu_i \geq 0, \; i = 1, \ldots, m \right\}$$

defined by a finite set of points y^i which are preferred to y, i.e. the cone $\mathcal{D}(y)$ depends on the point y.

Property (1.3) was related to the transitivity of a binary relation by Chew who shortly discussed in [32] the binary relation defined in (1.1) and who spoke of a domination structure if transitivity is given. Under the additional assumption that $\mathcal{D}(y)$ are pointed convex cones for all y Chew speaks of a normal domination structure. The discussion of the properties of the binary relations as presented here was first provided in [46]. Property (1.8) is similar to the f-inclusive condition defined by Huang et al. in [87]. Example 1.12 is a modification of an example presented there.

In [120, Definition 5.55] and [121, Definition 1], Mordukhovich defines a general preference relation by a binary relation $R \subset Y \times Y$ which satisfies several assumptions including the so-called *almost transitivity* which in particular says for elements $u, v, z \in Y$ that if u is preferred to z and v is preferred to u, then v is also preferred to z. This transitivity holds for the variable ordering structures, as presented here, only under additional assumptions on $\mathcal{D}$ which we do not assume in general, compare Lemma 1.10. However, Mordukhovich mentions in [120–122] that the almost transitivity requirement may be violated for some natural preferences which are significant in applications.

Bishop-Phelps cones have been introduced by Bishop and Phelps in 1962 in [21] as a class of ordering cones that have a rich mathematical structure. Some of the examples for these cones as well as the collection of properties of BP cones is due to

Jahn [93]. Results concerning BP cones w.r.t. a continuous linear functional $\phi \in Y^*$ with $\|\phi\|_* = 1$ can be found in the paper by Ha and Jahn [79]. The results on the representability of cones as BP cones are due to Petschke [129]. Bishop-Phelps cones in vector optimization have also gained special attention by Bednarczuk and Przybyla in [13]. Augmented dual cones were introduced by Kasimbeyli [100] who also provides in [100, Example 4.7] a representation of the cone $(\mathbb{R}^n_+)^{a*}$. The definition of supernormal cones originates from Isac [90] who defined them under the name nuclear cone in 1983. The equivalence to BP cones was stated by Isac and Bahya in [91]. Variable ordering structures given by Bishop-Phelps cones were, in this general form, introduced in [52]. Some of the examples for ordering maps are based on [48, 52].

The mentioned applications are based on the collection given in [46, 47], see also [50]. The application of a variable ordering structure in medical image registration (Sect. 1.3.1) is taken from the diploma thesis of Wacker [146], see also the publication [147] on the same topic. See [60] for a short introduction to image registration. The notion of equitability was related to a variable ordering structure by Baatar and Wiecek in [9].

Chapter 2
Optimality Concepts and Their Characterization

A variable ordering defined on the real linear space Y induces by the relation $\leq_1$ (given in (1.1)) and by the relation $\leq_2$ (given in (1.2)) an optimality concept which will be denoted as nondominated and as minimal elements, respectively. These two, fundamentally different concepts coincide only in the case of a non-variable ordering, i.e. if $\mathcal{D}(y) = K$ for all $y \in Y$ and some pointed convex cone K. Also weaker and stronger concepts can be derived as weakly or strongly nondominated/minimal elements or properly nondominated/minimal elements. In this chapter we first recall the optimality concepts from partially ordered spaces and then we introduce the various optimality notions for vector optimization problems with variable ordering structures. In addition to that we collect some basic properties of these notions.

We start by discussing optimal elements of a set. Having these concepts available directly allows us to define optimal solutions of a vector optimization problem

$$\min_{x \in S} f(x)$$

for some vector-valued map $f : X \to Y$ with X, Y real linear spaces and $S \subset X$ a nonempty subset: one only needs to determine the optimal elements of the set $f(S)$ and then to find their pre-images.

2.1 Optimality Concepts in Partially Ordered Spaces

In this section we recall the optimality concepts used in vector optimization in a partially ordered space. For that, we assume that Y is a real linear space which is partially ordered by a convex cone $K \subset Y$. Let A be a nonempty subset of Y. In partially ordered spaces we denote optimal elements as efficient elements according to the following definition.

G. Eichfelder, *Variable Ordering Structures in Vector Optimization*, Vector Optimization, 27
DOI 10.1007/978-3-642-54283-1_2, © Springer-Verlag Berlin Heidelberg 2014

Definition 2.1. An element $\bar{y} \in A$ is an *efficient element* of the set A if

$$(\{\bar{y}\} - K) \cap A \subset \{\bar{y}\} + K. \tag{2.1}$$

If K is additionally pointed, then (2.1) reduces to

$$(\{\bar{y}\} - K) \cap A = \{\bar{y}\}. \tag{2.2}$$

A partial ordering which is not antisymmetric, i.e. which is induced by a non-pointed ordering cone, is difficult to interpret. For that reason in most cases K is assumed to be pointed and thus, later, we also assume for the given variable ordering that the images of the ordering map are pointed convex cones. For handling also non-pointed cones, Borwein [24] replaced K in (2.2) with $\tilde{K} := (K \backslash (K \cap (-K))) \cup \{0_Y\}$. This could also be done for the images $\mathcal{D}(y)$ of an ordering map.

In the following, we always assume K to be a pointed convex cone.

For efficient elements in a partially ordered space, weaker (assuming $\mathrm{cor}(K) \neq \emptyset$) and stronger concepts are known.

Definition 2.2. (a) Assume $\mathrm{cor}(K) \neq \emptyset$. An element $\bar{y} \in A$ is a *weakly efficient* element of the set A, if

$$(\{\bar{y}\} - \mathrm{cor}(K)) \cap A = \emptyset.$$

(b) An element $\bar{y} \in A$ is a *strongly efficient* element of the set A, if

$$A \subset \{\bar{y}\} + K.$$

Weakly efficient elements are useful as they are completely characterized by linear scalarizations, but in applications one is usually not interested in only weakly efficient elements. Any efficient element of A is also a weakly efficient element (provided that K is pointed and $\mathrm{cor}(K) \neq \emptyset$), and any strongly efficient element of A is also an efficient element of A. If there is some strongly efficient element $\bar{y} \in A$, then it is the unique efficient element of the set. By replacing K by $-K$ in the above definitions, we obtain corresponding concepts of *(weakly/strongly) max-efficient* elements of a set A.

There are also other, stronger optimality notions, compared to efficiency, in vector optimization with a partially ordered space, known as properly efficient elements. We assume for all of these definitions for simplicity that $(Y, \|\cdot\|)$ is a real normed space and K is a closed pointed convex cone. Various types of these optimality concepts have been introduced. We give here the definitions in the sense of Henig, Benson and Borwein. For the definition of Borwein properly efficient elements, we need the concept of the contingent cone:

Definition 2.3. Let Ω be a nonempty subset of a real normed space $(Y, \|\cdot\|)$ and some $\bar{y} \in \mathrm{cl}(\Omega)$ be given. The set

$$T(\Omega, \bar{y}) := \{h \in Y \mid \exists (\lambda_n)_{n \in \mathbb{N}} \subset \mathbb{R}_{++}, \ \exists (y_n)_{n \in \mathbb{N}} \subset \Omega$$
$$\text{such that } \lim_{n \to \infty} y_n = \bar{y} \text{ and}$$
$$h = \lim_{n \to \infty} \lambda_n (y_n - \bar{y}) \}$$

is called *contingent cone* (or the *Bouligand tangent cone*) to Ω at $\bar{y}$.

Here, $\mathbb{R}_{++} = \{x \in \mathbb{R} \mid x > 0\}$.

In the convex case, contingent cones to some set Ω at $\bar{y}$ are related to the closure of the cone generated by the set $\Omega - \{\bar{y}\}$:

Lemma 2.4 ([94, Chap. 3]). *Let Ω be a nonempty subset of a real normed space $(Y, \| \cdot \|)$ and some $\bar{y} \in \Omega$ be given.*

(i) It holds

$$T(\Omega, \bar{y}) \subset cl(cone(\Omega - \{\bar{y}\})).$$

(ii) If Ω is starshaped w.r.t. $\bar{y}$, i.e. if for any $y \in \Omega$,

$$\lambda y + (1 - \lambda) \bar{y} \in \Omega \text{ for all } \lambda \in [0, 1],$$

then

$$T(\Omega, \bar{y}) = cl(cone(\Omega - \{\bar{y}\})).$$

Definition 2.5. (a) An element $\bar{y} \in A$ is a *properly efficient* element of A (in the sense of Henig [83]) if it is an efficient element of A and if there is a convex cone $C \subset Y$ with $K \setminus \{0_Y\} \subset int C$ such that $\bar{y}$ is an efficient element of A w.r.t. the partial ordering introduced by C, i.e.

$$\bar{y} \notin \{y\} + C \ \forall \ y \in A \setminus \{\bar{y}\}.$$

(b) An element $\bar{y} \in A$ is a *properly efficient* element of A (in the sense of Benson [16]) if it is an efficient element of A and if $\bar{y}$ is an efficient element of the set

$$\{\bar{y}\} + cl(cone(A + K - \{\bar{y}\})). \tag{2.3}$$

(c) An element $\bar{y} \in A$ is a *properly efficient* element of A (in the sense of Borwein [24]) if it is an efficient element of A and if $\bar{y}$ is an efficient element of the set

$$\{\bar{y}\} + T(A + K, \bar{y}). \tag{2.4}$$

The original definitions are slightly different but equivalent to the above ones. For instance, in (b), instead of requiring that $\bar{y}$ is an efficient element of the set defined in (2.3), i.e.

$$(\{\bar{y}\} - K) \cap (\{\bar{y}\} + cl(cone(A + K - \{\bar{y}\}))) = \{\bar{y}\},$$

in the original definition it is required that 0_Y is an efficient element of the set $\mathrm{cl}(\mathrm{cone}(A + K - \{\bar{y}\}))$, i.e. that

$$(-K) \cap (\mathrm{cl}(\mathrm{cone}(A + K - \{\bar{y}\}))) = \{0_Y\}.$$

The following example illustrates some of the optimality notions:

Example 2.6. Let $Y = \mathbb{R}^2$, $K = \mathbb{R}^2_+$ and

$$A := \{y \in \mathbb{R}^2 \mid \|y\|_2 \leq 1\} \cup \{y \in \mathbb{R}^2 \mid y_1 \geq 0,\ y_2 \geq -1\}.$$

Then all elements of the set

$$\{y \in \mathbb{R}^2 \mid \|y\|_2 = 1,\ y_1 \leq 0,\ y_2 \leq 0\}$$

are efficient elements of A w.r.t. $\mathcal{D}$. All elements of the set

$$\{y \in \mathbb{R}^2 \mid \|y\|_2 = 1,\ y_1 \leq 0,\ y_2 \leq 0\} \cup \{y \in \mathbb{R}^2 \mid y_1 \geq 0,\ y_2 = -1\}$$

are weakly efficient elements of A w.r.t. $\mathcal{D}$. There is no strongly efficient element of A. All elements of the set

$$\{y \in \mathbb{R}^2 \mid \|y\|_2 = 1,\ y_1 < 0,\ y_2 < 0\}$$

are properly efficient elements (in the sense of Henig/Benson/Borwein).

By Lemma 2.4, any properly efficient element in the sense of Benson in also properly efficient in the sense of Borwein. Moreover, if the set A is convex, then an element $\bar{y}$ is properly efficient in the sense of Benson if and only if it is properly efficient in the sense of Borwein.

2.2 Optimality Concepts for Variable Ordering Structures

In the following we assume Y to be a real linear space equipped with a variable ordering structure defined by the ordering map $\mathcal{D}: Y \to 2^Y$ with $\mathcal{D}(y)$ a pointed convex cone for all $y \in Y$. Additionally, let A be a nonempty subset of Y.

Based on the relation $\leq_1$ defined in (1.1) a candidate element is called a nondominated element if it is not dominated by other reference elements w.r.t. their cone. Instead, in view of the relation $\leq_2$ defined in (1.2), a candidate element is called a minimal element if it is not dominated by any other reference element w.r.t. the cone associated with the candidate element. These two notions are formally defined by:

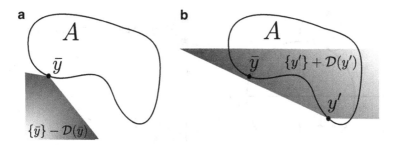

Fig. 2.1 (**a**) $\bar{y}$ is a minimal element of A w.r.t. $\mathcal{D}$. (**b**) $\bar{y}$ is not a nondominated element of A w.r.t. $\mathcal{D}$, cf. [46]

Definition 2.7. (a) An element $\bar{y} \in A$ is a *nondominated element* of the set A w.r.t. the ordering map $\mathcal{D}$, if $y \not\leq_1 \bar{y}$ for all $y \in A \setminus \{\bar{y}\}$, i.e. if no $y \in A$ exists such that

$$\bar{y} \in \{y\} + \mathcal{D}(y) \setminus \{0_Y\}, \tag{2.5}$$

or, equivalently,

$$\bar{y} \notin \bigcup_{y \in A} \{y\} + (\mathcal{D}(y) \setminus \{0_Y\}).$$

(b) An element $\bar{y} \in A$ is a *minimal element* of the set A w.r.t. the ordering map $\mathcal{D}$, if $y \not\leq_2 \bar{y}$ for all $y \in A \setminus \{\bar{y}\}$, i.e. if no $y \in A$ exists such that

$$\bar{y} \in \{y\} + \mathcal{D}(\bar{y}) \setminus \{0_Y\},$$

or, equivalently,

$$(\{\bar{y}\} - \mathcal{D}(\bar{y})) \cap A = \{\bar{y}\}. \tag{2.6}$$

For an illustration of both optimality concepts see Fig. 2.1.

Remark 2.8. For the definitions given in Definition 2.7 we do not require that the sets $\mathcal{D}(y)$ are pointed convex cones. Instead, $\mathcal{D}(y)$ can be an arbitrary set (with $0_Y \in \mathcal{D}(y)$) for any $y \in Y$.

For $\mathcal{D}(y) = K$ for all $y \in Y$ for some pointed convex cone K both definitions, Definition 2.7(a) and (b), coincide. I that case, K defines a partial ordering on Y and the minimal and the nondominated elements are exactly the efficient elements w.r.t. this partial ordering. Otherwise these notions are generally not related to each other in the sense that the one does not imply the other. This is illustrated with the following examples. We will show, however, in Chap. 7, that the notions are connected via duality in the sense that the nondominated elements of a set are the

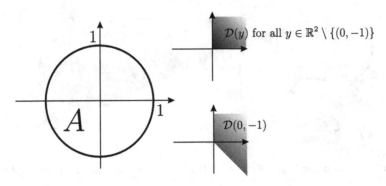

Fig. 2.2 Set A and the cones $\mathcal{D}(y)$ of Example 2.10, cf. [46]

minimal elements of some dual set (w.r.t. the "negative" of the variable ordering structure).

Example 2.9. Let Y be the Euclidean space $\mathbb{R}^2$,

$$A := \{y \in \mathbb{R}^2 \mid \|y\|_2 \leq 1\}$$

and let $\mathcal{D}: \mathbb{R}^2 \to 2^{\mathbb{R}^2}$ be defined by

$$\mathcal{D}(y) := \begin{cases} \mathbb{R}^2_+ & \forall y \in \mathbb{R}^2 \setminus \{(0,-1),(-1,0)\}, \\ \{(z_1,z_2) \in \mathbb{R}^2 \mid z_1 \leq 0,\ z_2 \geq 0\} & \text{for } y = (0,-1), \\ \{(z_1,z_2) \in \mathbb{R}^2 \mid z_1 \geq 0,\ z_2 \leq 0\} & \text{for } y = (-1,0). \end{cases}$$

Then all elements of the set $\{(y_1,y_2) \in \mathbb{R}^2 \mid \|y\|_2 = 1,\ y_1 \leq 0,\ y_2 \leq 0\}$ are minimal elements w.r.t. $\mathcal{D}$ but there is no nondominated element of A w.r.t. $\mathcal{D}$.

Example 2.10. Let Y and A be specified as in Example 2.9, i.e.

$$A := \{y \in \mathbb{R}^2 \mid \|y\|_2 \leq 1\},$$

and let $\mathcal{D}: \mathbb{R}^2 \to 2^{\mathbb{R}^2}$ be defined by

$$\mathcal{D}(y) := \begin{cases} \mathbb{R}^2_+ & \forall\ y \in \mathbb{R}^2 \setminus \{(0,-1)\}, \\ \{(z_1,z_2) \in \mathbb{R}^2 \mid z_1 + z_2 \geq 0, z_1 \geq 0\} & \text{for } y = (0,-1), \end{cases}$$

see Fig. 2.2. Then $(0,-1)$ is a nondominated element of A w.r.t. $\mathcal{D}$ but not a minimal element of A w.r.t. $\mathcal{D}$. The set of all nondominated elements of A w.r.t. $\mathcal{D}$ is $\{(y_1,y_2) \in \mathbb{R}^2 \mid y_1 \in [-1,0],\ y_2 = -\sqrt{1-y_1^2}\}$, and the set of all minimal elements of A w.r.t. $\mathcal{D}$ is $\{(y_1,y_2) \in \mathbb{R}^2 \mid y_1 \in [-1,0),\ y_2 = -\sqrt{1-y_1^2}\}$.

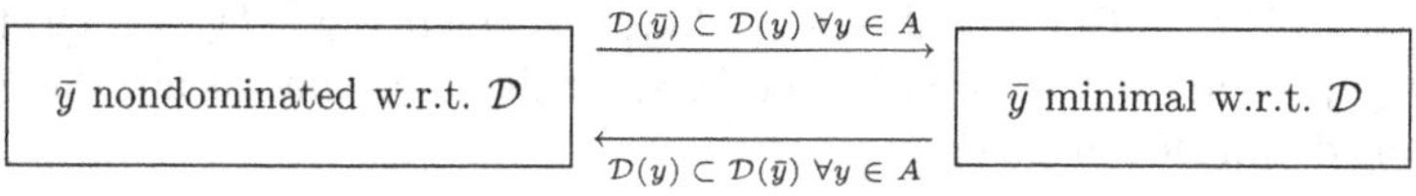

Fig. 2.3 Diagram illustrating the result of Lemma 2.11

Under strong assumptions on the ordering map $\mathcal{D}$ we obtain the following obvious relations between the ordering concepts, see also Fig. 2.3:

Lemma 2.11. *(i) If $\bar{y}$ is a minimal element of A w.r.t. $\mathcal{D}$ and $\mathcal{D}(y) \subset \mathcal{D}(\bar{y})$ for all $y \in A$, then $\bar{y}$ is also a nondominated element of A w.r.t. $\mathcal{D}$.*
(ii) If $\bar{y}$ is a nondominated element of A w.r.t. $\mathcal{D}$ and $\mathcal{D}(\bar{y}) \subset \mathcal{D}(y)$ for all $y \in A$, then $\bar{y}$ is also a minimal element of A w.r.t. $\mathcal{D}$.

For variable ordering structures we obtain analogously to Definition 2.2 the following weaker and stronger notions.

Definition 2.12. (a) Assume $\operatorname{cor}(\mathcal{D}(y)) \neq \emptyset$ for all $y \in A$. An element $\bar{y} \in A$ is a *weakly nondominated element* of the set A w.r.t. $\mathcal{D}$ if no $y \in A$ exists such that

$$\bar{y} \in \{y\} + \operatorname{cor}(\mathcal{D}(y)).$$

(b) An element $\bar{y} \in A$ is a *strongly nondominated element* of the set A w.r.t. $\mathcal{D}$ if

$$\bar{y} \in \{y\} - \mathcal{D}(y) \quad \text{for all} \ y \in A.$$

(c) Assume $\operatorname{cor}(\mathcal{D}(\bar{y})) \neq \emptyset$ for some element $\bar{y} \in A$. Then $\bar{y}$ is a *weakly minimal element* of the set A w.r.t. $\mathcal{D}$ if no $y \in A$ exists such that

$$\bar{y} \in \{y\} + \operatorname{cor}(\mathcal{D}(\bar{y})),$$

i.e. if

$$(\{\bar{y}\} - \operatorname{cor}(\mathcal{D}(\bar{y}))) \cap A = \emptyset.$$

(d) An element $\bar{y} \in A$ is a *strongly minimal element* of the set A w.r.t. $\mathcal{D}$ if

$$A \subset \{\bar{y}\} + \mathcal{D}(\bar{y}).$$

The concept of strongly minimal elements is a stronger notion compared to minimal elements w.r.t. $\mathcal{D}$: we do not only demand that $y \notin \{\bar{y}\} - \mathcal{D}(\bar{y})$ for all $y \in A \setminus \{\bar{y}\}$, but even that $y \in \{\bar{y}\} + \mathcal{D}(\bar{y})$ for all $y \in A \setminus \{\bar{y}\}$, i.e. there should be no other element which is in some sense "close" to being a preferred element. The same considerations hold for the concept of strongly nondominated elements w.r.t. $\mathcal{D}$. There, we do not only assume that the elements are nondominated elements w.r.t.

$\mathcal{D}$, i.e. that $\bar{y} \notin \{y\} + \mathcal{D}(y)$ for all $y \in A \setminus \{\bar{y}\}$, but even that $\bar{y} \in \{y\} - \mathcal{D}(y)$ for all $y \in A \setminus \{\bar{y}\}$, what can be interpreted in the sense that the elements are "far away" from being dominated by any other element.

We illustrate the weaker and stronger concepts by the following examples.

Example 2.13. Let Y be the Euclidean space $\mathbb{R}^2$,

$$A := \{(y_1, y_2) \in \mathbb{R}^2 \mid y_1 \geq 0, \ y_2 \geq 0, \ y_2 \geq 1 - y_1\}$$

and $\mathcal{D} \colon \mathbb{R}^2 \to 2^{\mathbb{R}^2}$ be defined by

$$\mathcal{D}(y_1, y_2) := \begin{cases} \text{cone conv } \{(y_1, y_2), (1, 0)\} & \text{if } (y_1, y_2) \in \mathbb{R}_+^2, \ y_2 \neq 0, \\ \mathbb{R}_+^2 & \text{otherwise.} \end{cases}$$

One can check that $\{(y_1, y_2) \in A \mid y_1 + y_2 = 1\}$ is the set of all nondominated elements of A w.r.t. $\mathcal{D}$ and

$$\{(y_1, y_2) \in A \mid y_1 + y_2 = 1 \ \vee \ y_1 = 0 \ \vee \ y_2 = 0\}$$

is the set of all weakly nondominated elements of A w.r.t. $\mathcal{D}$. There is no strongly nondominated element of A w.r.t. $\mathcal{D}$.

Example 2.14. Let Y be the Euclidean space $\mathbb{R}^2$,

$$A := \{(y_1, y_2) \in \mathbb{R}^2 \mid y_1 \leq y_2 \leq 2y_1\}$$

and $\mathcal{D} \colon \mathbb{R}^2 \to 2^{\mathbb{R}^2}$ be defined by

$$\mathcal{D}(y_1, y_2) := \begin{cases} \mathbb{R}_+^2 & \text{if } y_2 = 0, \\ \text{cone conv } \{(y_1, |y_2|), (1, 0)\} & \text{otherwise.} \end{cases}$$

One can check that $(0, 0) \in A$ is a strongly minimal and also a strongly nondominated element of A w.r.t. $\mathcal{D}$.

Again, for $\mathcal{D}(y) = K$ for all $y \in Y$ with K a pointed convex cone (and $\operatorname{cor}(K) \neq \emptyset$ when needed) the above weak and strong concepts coincide with those in a partially ordered space. In the following we study the connections between the optimal elements w.r.t. a variable ordering structure and the optimal elements w.r.t. a partial ordering in more detail.

Lemma 2.15. *(i) An element $\bar{y}$ is a minimal element of A w.r.t. $\mathcal{D}$ if and only if it is an efficient element of A in the linear space Y partially ordered by the cone $K := \mathcal{D}(\bar{y})$.*

(ii) Let $\operatorname{cor}(\mathcal{D}(\bar{y})) \neq \emptyset$ for some $\bar{y} \in A$. The element $\bar{y}$ is a weakly minimal element of A w.r.t. $\mathcal{D}$ if and only if it is a weakly efficient element of A in the linear space Y partially ordered by the cone $K := \mathcal{D}(\bar{y})$.

Proof. The assertion follows directly from the definitions.   □

For nondominated elements w.r.t. a variable ordering we obtain the following sufficient conditions.

Lemma 2.16. *Let $\mathcal{D}(A)$ be convex and pointed.*

(i) *Any efficient element of A in the linear space Y partially ordered by the cone $K := \mathcal{D}(A)$ is also a nondominated element of A w.r.t $\mathcal{D}$.*

(ii) *Let $cor(\mathcal{D}(y)) \neq \emptyset$ for all $y \in A$. Any weakly efficient element of A in the linear space Y partially ordered by $K := \mathcal{D}(A)$ is also a weakly nondominated element of A w.r.t $\mathcal{D}$.*

Proof. (i) If $\bar{y}$ is efficient, then $(\{\bar{y}\} - \mathcal{D}(A)) \cap A = \{\bar{y}\}$, i.e. for any $y \in A \setminus \{\bar{y}\}$ it holds $y \notin \{\bar{y}\} - \mathcal{D}(A)$ and thus $\bar{y} \notin \{y\} + \mathcal{D}(y)$.

(ii) If $\bar{y}$ is weakly efficient, then $(\{\bar{y}\} - cor(\mathcal{D}(A))) \cap A = \emptyset$, i.e. for any $y \in A$ it holds $y \notin \{\bar{y}\} - cor(\mathcal{D}(A))$ and thus $\bar{y} \notin \{y\} + cor(\mathcal{D}(y))$.

□

Analogously, we can formulate necessary conditions for nondominated elements w.r.t. a variable ordering structure based on efficient elements. A slightly weaker result is given later in Lemma 2.37.

Lemma 2.17. *(i) Any nondominated element of A w.r.t. $\mathcal{D}$ is also an efficient element of A with the linear space Y partially ordered by $K := \bigcap_{y \in A} \mathcal{D}(y)$.*

(ii) *Let $cor(\mathcal{D}(y)) \neq \emptyset$ for all $y \in A$ and set $K := \bigcap_{y \in A} \mathcal{D}(y)$. If $cor(K) \neq \emptyset$ then any weakly nondominated element of A w.r.t. $\mathcal{D}$ is also a weakly efficient element of A with the linear space Y partially ordered by K.*

Proof. (i) $\bar{y}$ nondominated of A w.r.t. $\mathcal{D}$ is equivalent to

$$\bar{y} \notin \{y\} + \mathcal{D}(y) \setminus \{0_Y\} \text{ for all } y \in A,$$

and hence $\bar{y} \notin \{y\} + K \setminus \{0_Y\}$ for all $y \in A$ or

$$(\{\bar{y}\} - K) \cap A = \{\bar{y}\}.$$

(ii) $\bar{y}$ weakly nondominated of A w.r.t. $\mathcal{D}$ is equivalent to

$$\bar{y} \notin \{y\} + cor(\mathcal{D}(y)) \text{ for all } y \in A,$$

and hence $\bar{y} \notin \{y\} + cor(K)$ for all $y \in A$ or

$$(\{\bar{y}\} - cor(K)) \cap A = \emptyset.$$

□

We sum up the result of the Lemmas 2.15–2.17 together with some obvious relations in a diagram in Fig. 2.4.

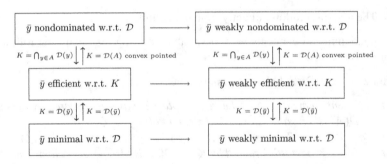

Fig. 2.4 Diagram illustrating the results of Lemmas 2.15–2.17 under the assumption $cor(\mathcal{D}(y)) \neq \emptyset$ and $cor(K) \neq \emptyset$ whenever considered

In the following theorem we extend Lemma 2.17 to a necessary condition which is, under some assumptions, also sufficient. We need the notion of external stability which is also denoted domination property, see [113] and the references therein.

Definition 2.18. Let the linear space Y be partially ordered by some convex cone K and let Ω be a nonempty subset of A. Then Ω is said to be *externally stable* if for all $y \in A \setminus \Omega$ there exists some $\bar{y} \in \Omega$ such that $y \in \{\bar{y}\} + K$.

Thus, external stability is satisfied if for all $y \in A$ there exists some $\bar{y} \in \Omega$ such that $y \in \{\bar{y}\} + K$, i.e. $A \subset \Omega + K$. In [134, Sect. 3.2] and [113] conditions ensuring the external stability of the set of efficient elements are given. Such a sufficient condition for the external stability of the set of efficient elements in a real linear space partially ordered by a convex cone K is given in the following.

Lemma 2.19 ([134, Theorem 3.2.10]). *Let Y be a real topological linear space, $K \subset Y$ a closed pointed convex cone and A a K-compact set, i.e. let the sets*

$$(\{y\} - K) \cap A$$

be compact for all $y \in A$. Let $\mathcal{E}_K$ denote the set of efficient elements of A w.r.t. the partial ordering introduced by K. Then $\mathcal{E}_K$ is externally stable, i.e. $A \subset \mathcal{E}_K + K$.

For instance, if A is compact then it is also K-compact for any closed cone K. In case A is an open set, the set of efficient elements $\mathcal{E}_K$ is empty and thus not an externally stable set of A. Next we give the announced theorem.

Theorem 2.20. *Let $K \subset \bigcap_{y \in A} \mathcal{D}(y)$ be a pointed convex cone and let $\mathcal{E}_K$ denote the set of efficient elements of A w.r.t. the partial ordering introduced by K. Let $\mathcal{E}_K$ be externally stable and let*

$$y^1 \in \{y^2\} + K \text{ imply } \mathcal{D}(y^1) \subset \mathcal{D}(y^2) \text{ for all } y^1, y^2 \in A. \qquad (2.7)$$

Then $\bar{y} \in A$ is a nondominated element of A w.r.t. $\mathcal{D}$ if and only if $\bar{y} \in \mathcal{E}_K$ and $\bar{y}$ is a nondominated element of $\mathcal{E}_K$ w.r.t. $\mathcal{D}$.

Proof. By Lemma 2.17(i) and as $\mathcal{E}_K \subset A$, the condition is necessary.

To show that it is also sufficient, assume $\bar{y}$ is in $\mathcal{E}_K$ and a nondominated element of $\mathcal{E}_K$ w.r.t. $\mathcal{D}$ but not of A. Then there exists some $y \in A \setminus \mathcal{E}_K$ with $\bar{y} \in \{y\} + \mathcal{D}(y) \setminus \{0_Y\}$. As $y \in A \setminus \mathcal{E}_K$ and $\mathcal{E}_K$ is externally stable, there exists some $\hat{y} \in \mathcal{E}_K$ with $y \in \{\hat{y}\} + K \setminus \{0_Y\}$. The condition (2.7) implies $\mathcal{D}(y) \subset \mathcal{D}(\hat{y})$ and we obtain

$$\bar{y} \in \{\hat{y}\} + K + \mathcal{D}(y)$$
$$\subset \{\hat{y}\} + \mathcal{D}(y)$$
$$\subset \{\hat{y}\} + \mathcal{D}(\hat{y})$$

and $\bar{y} \neq \hat{y} \in \mathcal{E}_K$ in contradiction to $\bar{y}$ a nondominated element of $\mathcal{E}_K$ w.r.t. $\mathcal{D}$. □

The following lemma shows that condition (2.7) is satisfied if the binary relation defined by the ordering map is transitive.

Lemma 2.21. *Let $\mathcal{D}(y)$ be algebraically closed for all $y \in Y$ and let $K \subset \bigcap_{y \in A} \mathcal{D}(y)$. If the binary relation $\leq_1$ defined in (1.1) is transitive, then (2.7) is satisfied, i.e. $y^1 \in \{y^2\} + K$ implies $\mathcal{D}(y^1) \subset \mathcal{D}(y^2)$ for all $y^1, y^2 \in A$.*

Proof. The relation $\leq_1$ is transitive according to Lemma 1.10(ii) if and only if

$$\mathcal{D}(y + d) \subset \mathcal{D}(y) \text{ for all } y \in Y \text{ and for all } d \in \mathcal{D}(y).$$

For $y^1 \in \{y^2\} + K$ it holds $y^1 = y^2 + d$ with $d \in K \subset \mathcal{D}(y^2)$ and thus $\mathcal{D}(y^1) = \mathcal{D}(y^2 + d) \subset \mathcal{D}(y^2)$. □

We obtain a similar result for minimal elements w.r.t a variable ordering structure. However, note that we need no condition like (2.7) or any other transitivity-related assumption.

Theorem 2.22. *Let $K \subset \bigcap_{y \in A} \mathcal{D}(y)$ be a pointed convex cone and let $\mathcal{E}_K$ denote the set of efficient elements of A w.r.t. the partial ordering introduced by K and let $\mathcal{E}_K$ be externally stable. Then $\bar{y} \in A$ is a minimal element of A w.r.t. $\mathcal{D}$ if and only if $\bar{y} \in \mathcal{E}_K$ and $\bar{y}$ is a minimal element of $\mathcal{E}_K$ w.r.t. $\mathcal{D}$.*

Proof. By Lemma 2.15, the definition of an efficient element and the fact that $K \subset \mathcal{D}(y)$ for all $y \in A$ and as $\mathcal{E}_K \subset A$, the condition is necessary.

To show that the condition is also sufficient, assume $\bar{y}$ is in $\mathcal{E}_K$ and a minimal element of $\mathcal{E}_K$ w.r.t. $\mathcal{D}$ but not of A. Then there exists some $y \in A \setminus \mathcal{E}_K$ with $\bar{y} \in \{y\} + \mathcal{D}(\bar{y}) \setminus \{0_Y\}$. As $\mathcal{E}_K$ is externally stable, there exists some $\hat{y} \in \mathcal{E}_K$ with $y \in \{\hat{y}\} + K \setminus \{0_Y\}$. Then

$$\bar{y} \in \{\hat{y}\} + K + \mathcal{D}(\bar{y})$$
$$\subset \{\hat{y}\} + \mathcal{D}(\bar{y}),$$

and $\bar{y} \neq \hat{y} \in \mathcal{E}_K$ in contradiction to $\bar{y}$ a minimal element of $\mathcal{E}_K$ w.r.t. $\mathcal{D}$. □

Next, we collect some results on weakly and strongly optimal elements w.r.t. a variable ordering structure.

Lemma 2.23. *(i) Any strongly nondominated element of A w.r.t. $\mathcal{D}$ is also a nondominated element of A w.r.t. $\mathcal{D}$. Any strongly minimal element of A w.r.t. $\mathcal{D}$ is also a minimal element of A w.r.t. $\mathcal{D}$.*

(ii) Let $cor(\mathcal{D}(y)) \neq \emptyset$ for all $y \in A$. Then any nondominated element of A w.r.t. $\mathcal{D}$ is also a weakly nondominated element of A w.r.t. $\mathcal{D}$. Any minimal element of A w.r.t. $\mathcal{D}$ is also a weakly minimal element of A w.r.t. $\mathcal{D}$.

(iii) If $\bar{y}$ is a strongly minimal element of A w.r.t. $\mathcal{D}$ and if $\mathcal{D}(\bar{y}) \subset \mathcal{D}(y)$ for all $y \in A$, then $\bar{y}$ is also a strongly nondominated element of A.

(iv) If $\bar{y}$ is a strongly nondominated element of A w.r.t. $\mathcal{D}$, then the set of minimal elements of A w.r.t. $\mathcal{D}$ is empty or equals $\{\bar{y}\}$. If additionally $\mathcal{D}(A) = \bigcup_{y \in A} \mathcal{D}(y)$ is pointed, then $\bar{y}$ is the unique minimal element of A w.r.t. $\mathcal{D}$.

(v) If $\mathcal{D}(A)$ is pointed, then there is at most one strongly nondominated element of A w.r.t. $\mathcal{D}$.

Proof. (i)–(iii) follow directly from the definitions.

(iv) As $\bar{y}$ is a strongly nondominated element of A w.r.t. $\mathcal{D}$,

$$\bar{y} \in (\{y\} - \mathcal{D}(y)) \cap A \text{ for all } y \in A,$$

i.e. all elements in $A \setminus \{\bar{y}\}$ are not minimal elements of A w.r.t. $\mathcal{D}$. If $\mathcal{D}(A)$ is pointed, for any $y \in A$ the assumption $y \in \{\bar{y}\} - \mathcal{D}(\bar{y})$ together with $y \in \{\bar{y}\} + \mathcal{D}(y)$ (as $\bar{y}$ is strongly nondominated) implies

$$y \in (\{\bar{y}\} - \mathcal{D}(A)) \cap (\{\bar{y}\} + \mathcal{D}(A)) ,$$

i.e. $y = \bar{y}$ and thus $(\{\bar{y}\} - \mathcal{D}(\bar{y})) \cap A = \{\bar{y}\}$.

(v) Let $\bar{y}$ be a strongly nondominated element of A w.r.t. $\mathcal{D}$. If $\mathcal{D}(A)$ is pointed, then $\bar{y} - y \in -\mathcal{D}(y) \subset -\mathcal{D}(A)$ implies $\bar{y} - y \notin \mathcal{D}(A)$ for all $y \in A \setminus \{\bar{y}\}$, i.e. $y \notin \{\bar{y}\} - \mathcal{D}(\bar{y})$ for all $y \in A \setminus \{\bar{y}\}$ and thus no other element of A can be strongly nondominated w.r.t. $\mathcal{D}$.   □

For a diagram illustrating the relations between the notions of (weakly, strongly) nondominated and minimal elements see Fig. 2.5.

The following example illustrates that the assumption that $\mathcal{D}(A)$ is pointed is necessary for the conclusion in Lemma 2.23(iv).

Example 2.24. Let Y be the Euclidean space $\mathbb{R}^2$, $A := \mathbb{R}^2_+$ and $\mathcal{D} : \mathbb{R}^2 \to 2^{\mathbb{R}^2}$ be defined by

$$\mathcal{D}(y_1, y_2) := \begin{cases} \text{cone conv } \{(y_1, y_2), (1, 0)\} & \text{if } (y_1, y_2) \in \mathbb{R}^2_+ \setminus \{(0, 0)\}, \\ \text{cone conv } \{(1, 0), (-1, -1)\} & \text{if } (y_1, y_2) = (0, 0), \\ \mathbb{R}^2_+ & \text{otherwise.} \end{cases}$$

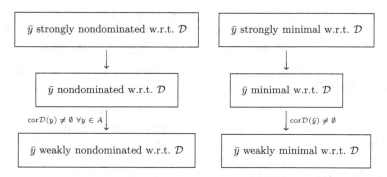

Fig. 2.5 Diagram illustrating the relation between the notions of (weakly, strongly) nondominated and minimal elements

Then $(0,0)$ is a strongly nondominated element of A w.r.t. $\mathcal{D}$ but not a minimal element of A w.r.t. $\mathcal{D}$. The cone $\mathcal{D}(A)$ is not pointed.

By replacing $\mathcal{D}$ by $\tilde{\mathcal{D}}$ with $\tilde{\mathcal{D}}(y) := -\mathcal{D}(y)$ for all $y \in Y$ in the Definitions 2.7 and 2.12, we obtain corresponding concepts of *(weakly/strongly) max-nondominated* and *(weakly/strongly) maximal elements* of a set A w.r.t. the ordering map $\mathcal{D}$.:

Definition 2.25. (a)   An element $\bar{y} \in A$ is a *max-nondominated element* of the set A w.r.t. the ordering map $\mathcal{D}$, if there is no $y \in A$ such that

$$\bar{y} \in \{y\} - \mathcal{D}(y) \setminus \{0_Y\}.$$

(b)   Assume $\mathrm{cor}(\mathcal{D}(y)) \neq \emptyset$ for all $y \in A$. An element $\bar{y} \in A$ is a *weakly max-nondominated element* of the set A w.r.t. the ordering map $\mathcal{D}$, if there is no $y \in A$ such that

$$\bar{y} \in \{y\} - \mathrm{cor}(\mathcal{D}(y)).$$

Definition 2.26. (a)   An element $\bar{y} \in A$ is a *maximal element* of the set A w.r.t. the ordering map $\mathcal{D}$, if

$$(\{\bar{y}\} + \mathcal{D}(\bar{y})) \cap A = \{\bar{y}\}.$$

(b)   Assume $\mathrm{cor}(\mathcal{D}(\bar{y})) \neq \emptyset$ for some element $\bar{y} \in A$. Then $\bar{y}$ is a *weakly maximal element* of the set A w.r.t. the ordering map $\mathcal{D}$, if

$$(\{\bar{y}\} + \mathrm{cor}(\mathcal{D}(\bar{y}))) \cap A = \emptyset.$$

Finally, we also give various notions of properly nondominated and properly efficient elements which generalize the definitions introduced in partially ordered real linear spaces to vector optimization with variable ordering structures.

Definition 2.27. Let $(Y, \| \cdot \|)$ be a real normed space and let $\mathcal{D}(y)$ be additionally closed for all $y \in Y$.

(a) An element $\bar{y} \in A$ is a *properly nondominated element* of A w.r.t. the ordering map $\mathcal{D}$ in the sense of Henig if it is a nondominated element of A w.r.t. the ordering map $\mathcal{D}$ and if there is a cone-valued map $\mathcal{K}: Y \to 2^Y$ with $\mathcal{K}(y)$ a convex cone for all $y \in Y$ and with $\mathcal{D}(y) \setminus \{0_Y\} \subset \operatorname{int}(\mathcal{K}(y))$ for all $y \in Y$ such that $\bar{y}$ is a nondominated element of A w.r.t. $\mathcal{K}$, i.e.

$$\bar{y} \notin \{y\} + \mathcal{K}(y) \ \forall \ y \in A \setminus \{\bar{y}\}.$$

(b) An element $\bar{y} \in A$ is a *properly nondominated element* of A w.r.t. the ordering map $\mathcal{D}$ in the sense of Benson if it is a nondominated element of A w.r.t. the ordering map $\mathcal{D}$ and if $\bar{y}$ is a nondominated element of the set

$$\{\bar{y}\} + \operatorname{cl}(\operatorname{cone}(\bigcup_{a \in A} (\{a\} + \mathcal{D}(a)) - \{\bar{y}\})).$$

(c) An element $\bar{y} \in A$ is a *properly nondominated element* of A w.r.t. the ordering map $\mathcal{D}$ in the sense of Borwein if it is a nondominated element of A w.r.t. the ordering map $\mathcal{D}$ and if $\bar{y}$ is a nondominated element of the set

$$\{\bar{y}\} + T(\bigcup_{a \in A} (\{a\} + \mathcal{D}(a)), \bar{y}).$$

(d) An element $\bar{y} \in A$ is a *properly minimal element* of A w.r.t. the ordering map $\mathcal{D}$ in the sense of Henig if it is a minimal element of A w.r.t. the ordering map $\mathcal{D}$ and if there is a cone-valued map $\mathcal{K}: Y \to 2^Y$ with $\mathcal{K}(y)$ a convex cone for all $y \in Y$ and with $\mathcal{D}(y) \setminus \{0_Y\} \subset \operatorname{int}(\mathcal{K}(y))$ for all $y \in Y$ such that $\bar{y}$ is a minimal element of A w.r.t. $\mathcal{K}$, i.e.

$$y \notin \{\bar{y}\} - \mathcal{K}(\bar{y}) \ \forall \ y \in A \setminus \{\bar{y}\}.$$

(e) An element $\bar{y} \in A$ is a *properly minimal element* of A w.r.t. the ordering map $\mathcal{D}$ in the sense of Benson if it is a minimal element of A w.r.t. the ordering map $\mathcal{D}$ and if $\bar{y}$ is a minimal element of the set

$$\{\bar{y}\} + \operatorname{cl}(\operatorname{cone}(A + \mathcal{D}(\bar{y}) - \{\bar{y}\})).$$

(f) An element $\bar{y} \in A$ is a *properly minimal element* of A w.r.t. the ordering map $\mathcal{D}$ in the sense of Borwein if it is a minimal element of A w.r.t. the ordering map $\mathcal{D}$ and if $\bar{y}$ is a minimal element of the set

$$\{\bar{y}\} + T(A + \mathcal{D}(\bar{y}), \bar{y}).$$

If $\mathcal{D}(y) = K$ for all $y \in Y$ then the definitions of a properly nondominated element w.r.t. $\mathcal{D}$ and of a properly minimal element w.r.t. $\mathcal{D}$ coincide with the concepts of a properly efficient element in a partially ordered space ordered by the closed pointed convex cone K, in the sense of Henig/Benson/Borwein. The following example illustrates some of the above notions:

Example 2.28. Let $A := \{y \in \mathbb{R}^2 \mid \|y\|_2 \le 1\}$ be the unit ball in the Euclidean space $Y = \mathbb{R}^2$ and let an ordering map $\mathcal{D}$ be defined by

$$\mathcal{D}(y) = \begin{cases} \{z \in \mathbb{R}^2 \mid z_1 \le 0, \ z_2 \ge 0\} & \text{if } y \in \{(0,1),(2,0)\}, \\ \mathbb{R}_+^2 & \text{else.} \end{cases}$$

The point $\bar{y} = (-1/\sqrt{2}, -1/\sqrt{2})$ is a properly nondominated and also a properly minimal point in the sense of Borwein.

Lemma 2.29. *Let $(Y, \| \cdot \|)$ be a real normed space and let $\mathcal{D}(y)$ be additionally closed for all $y \in Y$. The element $\bar{y}$ is a properly minimal element of A w.r.t. $\mathcal{D}$ in the sense of Henig if and only if it is a minimal element of A w.r.t. $\mathcal{D}$ and if there is a convex cone K with $\mathcal{D}(\bar{y}) \setminus \{0_Y\} \subset int(K)$ such that $y \notin \{\bar{y}\} - K$ for all $y \in A \setminus \{\bar{y}\}$.*

Proof. The only-if-part is immediate from Definition 2.27(d) by setting $K := \mathcal{K}(\bar{y})$. For the if-part, the set-valued map $\mathcal{K}$ can be defined by

$$\mathcal{K}(y) := \begin{cases} Y & \text{if } y \ne \bar{y}, \\ K & \text{if } y = \bar{y}. \end{cases}$$

Then $\mathcal{D}(y) \setminus \{0_Y\} \subset int(\mathcal{K}(y))$ for all $y \in Y$ and we are done. □

The next lemma follows directly from the definitions.

Lemma 2.30. *Let $(Y, \| \cdot \|)$ be a real normed space and let $\mathcal{D}(y)$ be additionally closed for all $y \in Y$. $\bar{y}$ is a properly minimal element of A in the sense of Henig/Benson/Borwein w.r.t. $\mathcal{D}$ if and only if it is a properly efficient element in the sense of Henig/Benson/Borwein of A with Y partially ordered by the closed convex cone $K := \mathcal{D}(\bar{y})$.*

Remark 2.31. As a consequence of the above Lemma and [75, Theorem 4.2], if $\bar{y}$ is a properly minimal element in the sense of Henig with $\mathcal{K}(\bar{y})$ pointed, then $\bar{y}$ is also a properly minimal element in the sense of Benson. As a consequence of the above Lemma and [100, Theorem 5.2, Remark 5.3], if $\bar{y}$ is a properly minimal element in the sense of Benson and if $\mathcal{D}(\bar{y})$ has a weakly compact base, then $\bar{y}$ is also a properly minimal element in the sense of Henig w.r.t. $\mathcal{D}$.

We also have the following relations between the proper optimality notions:

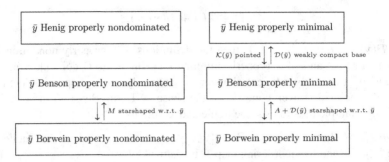

Fig. 2.6 Diagram illustrating the results of Remarks 2.31 and 2.32 with $M := \bigcup_{a \in A}(\{a\} + \mathcal{D}(a))$, cf. [55]

Remark 2.32.   (i) By Lemma 2.4, any Benson properly nondominated/ minimal element of A w.r.t. $\mathcal{D}$ is also a Borwein properly nondominated/minimal element of A w.r.t. $\mathcal{D}$.

(ii) If the set $M := \bigcup_{a \in A}(\{a\} + \mathcal{D}(a))$ is starshaped w.r.t. some element $\bar{y} \in A$, then by Lemma 2.4 the element $\bar{y}$ is a Benson properly nondominated element if and only if it is a Borwein properly nondominated element.

(iii) Also by Lemma 2.4, if $A + \mathcal{D}(\bar{y})$ is starshaped w.r.t. $\bar{y} \in A$, then the element $\bar{y}$ is a Benson properly minimal element if and only if it is a Borwein properly minimal element.

(iv) If the cones $\mathcal{D}(y)$ are closed, pointed, nontrivial and convex cones in $\mathbb{R}^m$, then by [134, Theorem 3.1.2], [75, Theorem 4.2], Henig's proper minimality is equivalent to Benson's proper minimality.

We sum up the results of the Remarks 2.32 and 2.31 in a diagram in Fig. 2.6.

The remaining relations between Henig properly nondominated and Benson properly nondominated are not as straightforward and are discussed in [55]: Henig properly nondominated implies in general not Benson properly nondominated. In case $\mathcal{D}(y)$ has a weakly compact base for each $y \in Y$, then there is an implication from Benson to Henig proper nondominatedness.

In addition to considering optimal elements of a set w.r.t. a variable ordering structure, all concepts apply also to an optimization problem with a vector-valued (or set-valued) objective map and with the objective space equipped with a variable ordering structure. For that assume that X and Y are real linear spaces and Y is equipped with a variable ordering structure defined by the cone-valued map $\mathcal{D}: Y \to 2^Y$ with $\mathcal{D}(y)$ a pointed convex cone for all $y \in Y$. Let $F : X \to 2^Y$ be a set-valued map and $S \subset X$ a nonempty set. We consider the following optimization problem:

$$\text{Minimize } F(x) \text{ subject to } x \in S. \qquad (\text{VP})$$

In case $F = f$ is a single-valued map $f : X \to Y$ we denote problem (VP) as a vector optimization problem. Otherwise we speak of a set optimization problem. Further, we denote the image of S under F with $F(S) = \bigcup_{x \in S} F(x)$.

The various notions of nondominated and minimal elements w.r.t. the ordering map $\mathcal{D}$ for sets naturally induce corresponding notions of solutions to the optimization problem (VP) as follows.

Definition 2.33. Let $\bar{x} \in X$ and $\bar{y} \in F(\bar{x})$.

(a) The pair $(\bar{x}, \bar{y})$ is a *globally "N" solution* of the problem (VP) w.r.t. the ordering map $\mathcal{D}$, if $\bar{y}$ is an "N" element of the set $F(S)$. Here, "N" may be (weakly/strongly) nondominated or max-nondominated, (weakly/strongly) minimal or maximal.

(b) The pair $(\bar{x}, \bar{y})$ is a *locally "N" solution* of the problem (VP) w.r.t. the ordering map $\mathcal{D}$, if $\bar{y}$ is an "N" element of the set $F(S \cap U_{\bar{x}})$, where $U_{\bar{x}}$ is some neighborhood of $\bar{x}$. Here, "N" may be (weakly/strongly) nondominated or max-nondominated or (weakly, strongly) minimal or maximal.

Analogously to the above definition one can define (weakly/strongly) efficient solutions of the problem (VP) w.r.t. a partial ordering introduced by a convex cone K. The same for all notions of proper optimality.

When no confusion occurs, we omit "globally" in the above definition and when $F = f$ is a single-valued map $f : X \to Y$, we put $\bar{y} = f(\bar{x})$ in Definition 2.33 and say $\bar{x}$ is a (globally) "N" solution.

In Definition 2.33 we define optimal solutions of a set optimization problem by the so-called vector approach. This approach is widespread, but might not be suitable for some applications. See Sect. 2.5 for a short discussion on this topic.

2.3 Characterization of Nondominated Elements

We examine in this section basic properties of nondominated elements of a set A w.r.t. a variable ordering structure. In the following we again assume Y to be a real linear space equipped with a variable ordering structure defined by the ordering map $\mathcal{D} : Y \to 2^Y$ with $\mathcal{D}(y)$ a pointed convex cone for all $y \in Y$. Additionally, let A be a nonempty subset of Y. Under rather weak assumptions, all weakly nondominated elements are a subset of the algebraic boundary ∂A of the set A.

Lemma 2.34. *(i) Let $cor(\mathcal{D}(y)) \neq \emptyset$ for all $y \in A$. If*

$$\bigcap_{y \in A} cor(\mathcal{D}(y)) \neq \emptyset$$

and $\bar{y} \in A$ is a weakly nondominated element of A w.r.t. $\mathcal{D}$, then $\bar{y} \in \partial A$.

(ii) If $\bigcap_{y \in A} \mathcal{D}(y) \neq \{0_Y\}$ and $\bar{y} \in A$ is a nondominated element of A w.r.t. $\mathcal{D}$, then $\bar{y} \in \partial A$.

Proof. (i) We assume that $\bar{y} \in \text{cor}(A)$. Let $d \in \bigcap_{y \in A} \text{cor}(\mathcal{D}(y))$. Then $d \neq 0_Y$ and there exists $\lambda > 0$ with $\bar{y} - \lambda d \in A \setminus \{\bar{y}\}$. As

$$-\lambda d \in -\bigcap_{y \in A} \text{cor}(\mathcal{D}(y)) \subset -\text{cor}(\mathcal{D}(\bar{y} - \lambda d))$$

we have $\bar{y} - \lambda d \in A \cap (\{\bar{y}\} - \text{cor}(\mathcal{D}(\bar{y} - \lambda d)))$ or

$$\bar{y} \in \{\bar{y} - \lambda d\} + \text{cor}(\mathcal{D}(\bar{y} - \lambda d)),$$

being a contradiction to $\bar{y}$ weakly nondominated. As also $\bar{y} \in A$ and thus $\bar{y} \notin \text{cor}(Y \setminus A)$ we get $\bar{y} \in \partial A$

(ii) We assume that $\bar{y} \in \text{cor}(A)$. Let $d \in \left(\bigcap_{y \in A} \mathcal{D}(y)\right) \setminus \{0_Y\}$. Then there exists $\lambda > 0$ with $\bar{y} - \lambda d \in A \setminus \{\bar{y}\}$. As

$$-\lambda d \in -\bigcap_{y \in A} \mathcal{D}(y) \subset -\mathcal{D}(\bar{y} - \lambda d)$$

we have $\bar{y} - \lambda d \in A \cap (\{\bar{y}\} - \mathcal{D}(\bar{y} - \lambda d))$ or $\bar{y} \in \{\bar{y} - \lambda d\} + \mathcal{D}(\bar{y} - \lambda d)$, being a contradiction to $\bar{y}$ nondominated. As $\bar{y} \in A$ we get $\bar{y} \in \partial A$. $\square$

The following example demonstrates that the assumption in Lemma 2.34(i) is necessary.

Example 2.35. Let Y be the Euclidean space $\mathbb{R}^2$, $A = [1, 3] \times [1, 3]$ and the ordering map $\mathcal{D} \colon \mathbb{R}^2 \to 2^{\mathbb{R}^2}$ be defined by

$$\mathcal{D}(y_1, y_2) := \begin{cases} \mathbb{R}_+^2 & \text{if } y_1 \geq 2, \\ \{(z_1, z_2) \in \mathbb{R}^2 \mid z_1 \leq 0, z_2 \geq 0\} & \text{otherwise.} \end{cases}$$

Then $\bar{y} = (2, 2)$ is a weakly nondominated element of A w.r.t. $\mathcal{D}$ but $\bar{y} \notin \partial A$.

It is easy to see that $A_1 \subset A_2$ implies that an element $\bar{y} \in A_1$, which is a (weakly) nondominated element of A_2 w.r.t. $\mathcal{D}$ is also a (weakly) nondominated element of A_1 w.r.t. $\mathcal{D}$. The following remark states a similar result for two ordering maps $\mathcal{D}_1$ and $\mathcal{D}_2$ with the one including the other in the sense that $\mathcal{D}_1(y) \subset \mathcal{D}_2(y)$ for all $y \in A$.

Remark 2.36. Let $\mathcal{D}_1, \mathcal{D}_2 \colon Y \to 2^Y$ be cone-valued maps with $\mathcal{D}_1(y)$ and $\mathcal{D}_2(y)$ a pointed convex cone for all $y \in Y$. According to the Definitions 2.7 and 2.12, if $\mathcal{D}_1(y) \subset \mathcal{D}_2(y)$ for all $y \in A$, then any nondominated element of the set A w.r.t. the ordering map $\mathcal{D}_2$ is also a nondominated element of the set A w.r.t. the ordering map $\mathcal{D}_1$. If $\text{cor}(\mathcal{D}_1(y)) \neq \emptyset$ for all $y \in A$, then this also holds if we replace nondominated by weakly nondominated.

This leads to the following relation—which also directly follows from Lemmas 2.16 and 2.17—between nondominated elements w.r.t. a variable ordering structure and efficient elements in a partially ordered space. Of course, corresponding statement can be formulated for weakly efficient and nondominated elements.

Lemma 2.37. *(i) Let $\bar{y} \in A$ be a nondominated element of the set A w.r.t. the ordering map $\mathcal{D}$. Then for any convex cone K with $K \subset \mathcal{D}(y)$ for all $y \in A$, $\bar{y}$ is also an efficient element of the set A with the space Y partially ordered by K.*
(ii) Let $\bar{y}$ be an efficient element of the set A with the space Y partially ordered by some pointed convex cone K. If $\mathcal{D}(y) \subset K$ for all $y \in A$, then $\bar{y}$ is also a nondominated element of the set A w.r.t. the ordering map $\mathcal{D}$.

In partially ordered spaces ordered by some pointed convex cone K it is known [158, Lemma 4.1] that an element is an efficient element of a set A if and only if it is an efficient element of the set $A + K$. This can be generalized to variable ordering structures only under the additional assumption of the transitivity of the binary relation on the set A. The advantage of studying $A + K$ instead of A lies for instance in a better applicability of linear scalarization results if $A + K$ is convex while the set A is not. In that case A is called *convex-like*, see [59] or [1, 2].

Lemma 2.38. *Define the set*

$$M := \bigcup_{y \in A} \{y\} + \mathcal{D}(y).$$

(i) If $\bar{y} \in M$ is a nondominated element of M w.r.t. $\mathcal{D}$, then $\bar{y} \in A$ and $\bar{y}$ is also a nondominated element of A w.r.t. $\mathcal{D}$.
(ii) If $\bar{y} \in A$ is a nondominated element of A w.r.t. $\mathcal{D}$, and if

$$\mathcal{D}(y + d) \subset \mathcal{D}(y) \text{ for all } y \in A \text{ and for all } d \in \mathcal{D}(y), \qquad (2.8)$$

then $\bar{y}$ is a nondominated element of M w.r.t. $\mathcal{D}$.

Proof. (i) If $\bar{y} \in M \setminus A$ then $\bar{y} \in \{y\} + (\mathcal{D}(y) \setminus \{0_Y\})$ for some $y \in A \subset M$ in contradiction to $\bar{y}$ a nondominated element of M w.r.t. $\mathcal{D}$. Thus $\bar{y} \in A$. Due to $A \subset M$, $\bar{y}$ is then also a nondominated element of A w.r.t. $\mathcal{D}$.

(ii) We assume that $\bar{y}$ is a nondominated element of A w.r.t. $\mathcal{D}$ but not of M, i.e. there exist $y \in A$ and $d_y \in \mathcal{D}(y) \setminus \{0_Y\}$ with $\bar{y} \in \{y + d_y\} + (\mathcal{D}(y + d_y) \setminus \{0_Y\})$. As $\mathcal{D}(y)$ is a pointed convex cone this implies

$$\begin{aligned} \bar{y} &\in \{y\} + (\mathcal{D}(y) \setminus \{0_Y\}) + (\mathcal{D}(y + d_y) \setminus \{0_Y\}) \\ &\subset \{y\} + (\mathcal{D}(y) \setminus \{0_Y\}) + (\mathcal{D}(y) \setminus \{0_Y\}) \\ &\subset \{y\} + (\mathcal{D}(y) \setminus \{0_Y\}), \end{aligned}$$

in contradiction to $\bar{y}$ being a nondominated element of A w.r.t. $\mathcal{D}$.

$\square$

Fig. 2.7 Set M of
Example 2.39, cf. [46]

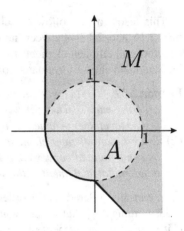

If the binary relation $\leq_1$ defined in (1.1) by the ordering map $\mathcal{D}$ is transitive, then condition (2.8) is satisfied, compare Lemma 1.10(ii).

In the following example all assumptions of Lemma 2.38(ii) are fulfilled, but $\mathcal{D}$ is not transitive on Y.

Example 2.39. Let Y, A and $\mathcal{D}$ be defined as in Example 2.10. Then

$$M = \{(y_1, y_2) \in \mathbb{R}^2 \mid y_1 \in [-1, 0], y_2 \geq -\sqrt{1 - y_1^2}\}$$

$$\cup \{(y_1, y_2) \in \mathbb{R}^2 \mid y_1 \geq 0, y_2 \geq -1 - y_1\},$$

see Fig. 2.7.

The set of nondominated elements of M w.r.t. $\mathcal{D}$ is

$$\left\{ (y_1, y_2) \in \mathbb{R}^2 \mid y_1 \in [-1, 0], \ y_2 = -\sqrt{1 - y_1^2} \right\}$$

and thus equals the set of nondominated elements of A w.r.t. $\mathcal{D}$. It is easy to verify that $\mathcal{D}(y + d) \subset \mathcal{D}(y)$ for all $y \in A$ and for all $d \in \mathcal{D}(y)$. However, the relation $\leq_1$ defined by the ordering map $\mathcal{D}$ is not transitive. For instance $(-1, -2) \leq_1 (0, -1)$ and $(0, -1) \leq_1 (2, -3)$ but $(-1, -2) \not\leq_1 (2, -3)$ because of $(2, -3) \notin \{(-1, -2)\} + \mathbb{R}_+^2$.

In general only the cones $\mathcal{D}(y)$ for $y \in A$ are of interest for modeling a decision making problem. Thus we have the freedom of setting $\mathcal{D}(y) := \{0_Y\}$ for all $y \in Y \setminus A$. This allows us to make the assumption (2.8) dispensable for the result in Lemma 2.38(ii):

Lemma 2.40. *Let* $\mathcal{D} \colon Y \to 2^Y$ *be given with* $\mathcal{D}(y) = \{0_Y\}$ *for all* $y \in Y \setminus A$ *and let* $M := \bigcup_{y \in A} \{y\} + \mathcal{D}(y)$. *Then an element* $\bar{y} \in Y$ *is a nondominated element of the set* A *w.r.t.* $\mathcal{D}$ *if and only if it is a nondominated element of the set* M *w.r.t.* $\mathcal{D}$.

Fig. 2.8 A section
$A_y = (\{y\} - K) \cap A$ of
some set A

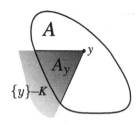

Proof. First assume $\bar{y}$ is a nondominated element of the set A w.r.t. $\mathcal{D}$. If it is not also nondominated of M w.r.t. $\mathcal{D}$ then there exists some $y \in A$ and some $d \in \mathcal{D}(y)$ such that

$$\bar{y} \in \{y + d\} + \mathcal{D}(y + d) \setminus \{0_Y\} \text{ with } y + d \notin A. \qquad (2.9)$$

Thus $y + d \in M \setminus A$ and $\mathcal{D}(y + d) = \{0_Y\}$ in contradiction to (2.9). The other implication follows from Lemma 2.38(i). □

For obtaining existence results for optimal elements in a partially ordered space under weak assumptions on the set, so-called sections are considered. If $K \subset Y$ denotes the ordering cone then to any $y \in Y$ with

$$A_y := (\{y\} - K) \cap A \neq \emptyset,$$

the set A_y is denoted a *section*, see Fig. 2.8. It holds that any (weakly) efficient element of a section is also a (weakly) efficient element of the set in the real linear space Y partially ordered by K, cf. [94, Lemma 6.2]. Using Lemma 2.16 we can extend this result to variable ordering structures.

Lemma 2.41. *Let $K := \mathcal{D}(A)$ be pointed and convex and let K thus introduce a partial ordering on Y. Consider for some $y \in Y$ the set $A_y := (\{y\} - K) \cap A$.*

(i) *Any efficient element of A_y w.r.t. K is also a nondominated element of A w.r.t. $\mathcal{D}$.*

(ii) *Let $cor(\mathcal{D}(y)) \neq \emptyset$ for all $y \in A$. Then any weakly efficient element of A_y w.r.t. K is also a weakly nondominated element of A w.r.t. $\mathcal{D}$.*

Proof. (i) According to [94, Lemma 6.2(a)], any efficient element of A_y is also an efficient element of A and thus by Lemma 2.16(i) also a nondominated element of A w.r.t. $\mathcal{D}$.

(ii) According to [94, Lemma 6.2(b)], any weakly efficient element of A_y is also a weakly efficient element of A and thus by Lemma 2.16(ii) also a weakly nondominated element of A w.r.t. $\mathcal{D}$. □

The result of Lemma 2.41 is useful for applying existence results for efficient elements as provided in [94, Chap. 6] also to nondominated elements—in case $\mathcal{D}(A)$

Fig. 2.9 Section A_y of
Example 2.43

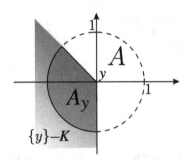

is pointed and convex. The following lemma directly relates nondominated elements
of a section w.r.t. an ordering map $\mathcal{D}$ with those of the set itself.

Lemma 2.42. *Let $K := \mathcal{D}(A)$ be convex and consider for some $y \in Y$ the set
$A_y = (\{y\} - K) \cap A$.*

*(i) Any nondominated element of A_y w.r.t. $\mathcal{D}$ is also a nondominated element of A
w.r.t. $\mathcal{D}$.*

*(ii) Let $cor(\mathcal{D}(y)) \neq \emptyset$ for all $y \in A$. Then any weakly nondominated element of
A_y w.r.t. $\mathcal{D}$ is also a weakly nondominated element of A w.r.t. $\mathcal{D}$.*

Proof. (i) Assume $\bar{y}$ is a nondominated element of A_y w.r.t. $\mathcal{D}$ but there exists
some $\hat{y} \in A \setminus A_y$ with $\bar{y} \in \{\hat{y}\} + \mathcal{D}(\hat{y})$. As $\bar{y} \in A_y$,

$$\hat{y} \in \{\bar{y}\} - \mathcal{D}(\hat{y}) \subset \{y\} - \mathcal{D}(A) - \mathcal{D}(\hat{y})$$
$$\subset \{y\} - (\mathcal{D}(A) + \mathcal{D}(A)) \subset \{y\} - \mathcal{D}(A),$$

in contradiction to $\hat{y} \notin A_y$.

(ii) Analogously to (i).

$\square$

Example 2.43. Let Y, A and $\mathcal{D}$ be defined as in Example 2.10, compare also
Example 2.39. Then $\mathcal{D}(A) = \{(z_1, z_2) \in \mathbb{R}^2 \mid z_1 + z_2 \geq 0,\ z_1 \geq 0\}$ is convex.
For instance, the section $A_y = (\{y\} - K) \cap A$, see Fig. 2.9, defined by $y = (0, 0)$
and $K = \mathcal{D}(A)$ contains all nondominated elements of A w.r.t. $\mathcal{D}$ and according
to Lemma 2.42 all nondominated elements of A_y w.r.t. $\mathcal{D}$ are also nondominated
elements of A w.r.t. $\mathcal{D}$.

We conclude this section by a result about the nondominated elements of the sum
of two sets.

Lemma 2.44. *Let A, $B \subset Y$ be nonempty subsets and let $\mathcal{D}$ satisfy*

$$\mathcal{D}(y^A) + \mathcal{D}(y^B) \subset \mathcal{D}(y^A + y^B) \qquad \text{for all } y^A \in A,\ y^B \in B. \qquad (2.10)$$

If $\bar{y} = \bar{y}^A + \bar{y}^B \in A + B$ with $\bar{y}^A \in A$, $\bar{y}^B \in B$ is a nondominated element of $A + B$ w.r.t. the ordering map $\mathcal{D}$, then $\bar{y}^A$ is a nondominated element of A w.r.t. the ordering map $\mathcal{D}$ and $\bar{y}^B$ is a nondominated element of B w.r.t. the ordering map $\mathcal{D}$.

Proof. We assume $\bar{y}^A$ is not a nondominated element of A w.r.t. the ordering map $\mathcal{D}$. Then there exists $y \in A$ with $\bar{y}^A \in \{y\} + (\mathcal{D}(y) \setminus \{0_Y\})$ and thus

$$\bar{y} = \bar{y}^A + \bar{y}^B \in \{y + \bar{y}^B\} + (\mathcal{D}(y) \setminus \{0_Y\}).$$

With $0_Y \in \mathcal{D}(\bar{y}^B)$ and because $\mathcal{D}$ satisfies (2.10) we get

$$\mathcal{D}(y) \subset \mathcal{D}(y) + \mathcal{D}(\bar{y}^B) \subset \mathcal{D}(y + \bar{y}^B).$$

Thus

$$\bar{y} \in \underbrace{\{y + \bar{y}^B\}}_{\in A+B} + (\mathcal{D}(y + \bar{y}^B) \setminus \{0_Y\})$$

in contradiction to $\bar{y}$ a nondominated element of $A + B$ w.r.t. the ordering map $\mathcal{D}$. The same for $\bar{y}^B$.   □

For a variable ordering structure satisfying (2.10) see the following example:

Example 2.45. Let $\mathcal{D}$ be specified as in Example 2.10. Let arbitrary sets A and B be given with $A, B \subset \mathbb{R}^2_+$. We get for any $y^A \in A$, $y^B \in B$ that $y_1^A + y_1^B \geq y_1^A$ and also $y_1^A + y_1^B \geq y_1^B$. Because $\mathcal{D}(y') \supset \mathcal{D}(y)$ for any $y, y' \in \mathbb{R}^2_+$ with $y_1' \geq y_1$, we conclude, using the convexity of $\mathcal{D}(y)$ for any $y \in \mathbb{R}^2$:

$$\mathcal{D}(y^A) + \mathcal{D}(y^B) \subset \mathcal{D}(y^A + y^B) + \mathcal{D}(y^A + y^B) = \mathcal{D}(y^A + y^B).$$

But note that on $\mathbb{R}^2_+$ the map $\mathcal{D}$ is constant.

If the set-valued map $\mathcal{D}$ is subadditive w.r.t. the cone $\{0_Y\}$ on Y, then condition (2.10) is satisfied. For the definition of subadditivity of a set-valued map see (3.9) in Chap. 3. However, we show in Lemma 3.23 that subadditivity is a too strong assumption for a cone-valued map as it implies that it is a constant-valued map.

The converse statement of Lemma 2.44 does not hold in general even not in the case of a partially ordered space [134, Remark 3.1.3].

2.4 Characterization of Minimal Elements

This section is devoted to basic properties of minimal elements of a set A w.r.t. a variable ordering structure. As before, let Y be a real linear space equipped with a variable ordering structure defined by the ordering map $\mathcal{D}: Y \to 2^Y$ with $\mathcal{D}(y)$ a pointed convex cone for all $y \in Y$. Additionally, let A be a nonempty subset of Y.

First we state that all (weakly) minimal elements are a subset of the algebraic boundary ∂A of the set A.

Lemma 2.46. *(i) Let $cor(\mathcal{D}(y)) \neq \emptyset$ for all $y \in A$. If $\bar{y} \in A$ is a weakly minimal element of A w.r.t. $\mathcal{D}$, then $\bar{y} \in \partial A$.*
(ii) If $\bar{y} \in A$ is a minimal element of A w.r.t. $\mathcal{D}$ and $\mathcal{D}(\bar{y}) \neq \{0_Y\}$, then $\bar{y} \in \partial A$.

Proof. (i) Follows directly from Lemma 2.15(ii) and the known fact that in partially ordered spaces all weakly efficient elements are a subset of the boundary of the set [42, Theorem 1.13]. However, it can also be easily shown by choosing any $d \in cor(\mathcal{D}(\bar{y}))$. Then the proof is analogous to the proof of Lemma 2.34(i).

(ii) Follows directly from Lemma 2.15(i) and the known fact that in partially ordered spaces all efficient elements are a subset of the boundary of the set [70, Theorem 2.9].

□

Again, $A_1 \subset A_2$ implies that an element $\bar{y} \in A_1$, which is a (weakly) minimal element of A_2 w.r.t. $\mathcal{D}$ is also a (weakly) minimal element of A_1 w.r.t. $\mathcal{D}$. The following remark relates minimal elements w.r.t. two ordering maps $\mathcal{D}_1$ and $\mathcal{D}_2$.

Remark 2.47. Let $\mathcal{D}_1, \mathcal{D}_2: Y \rightarrow 2^Y$ be cone-valued maps with $\mathcal{D}_1(y)$ and $\mathcal{D}_2(y)$ a pointed convex cone for all $y \in Y$. According to the Definitions 2.7 and 2.12, if $\bar{y}$ is a minimal element of the set A w.r.t. the ordering map $\mathcal{D}_2$ and $\mathcal{D}_1(\bar{y}) \subset \mathcal{D}_2(\bar{y})$, then $\bar{y}$ is also a minimal element of the set A w.r.t. the ordering map $\mathcal{D}_1$. If $cor(\mathcal{D}_1(\bar{y})) \neq \emptyset$, then this also holds if we replace minimal by weakly minimal.

We can relate the minimal elements of the set A with the minimal elements of the set $M = \bigcup_{y \in A} \{y\} + \mathcal{D}(y)$ only under strong assumptions on the variable ordering.

Lemma 2.48. *Define the set*

$$M := \bigcup_{y \in A} \{y\} + \mathcal{D}(y).$$

(i) If $\bar{y} \in A$ is a minimal element of M w.r.t. $\mathcal{D}$, then it is also a minimal element of A w.r.t. $\mathcal{D}$.
(ii) If $\bar{y} \in A$ is a minimal element of A w.r.t. $\mathcal{D}$ and if $\mathcal{D}(y) \subset \mathcal{D}(\bar{y})$ for all $y \in A$, then $\bar{y}$ is also a minimal element of M w.r.t $\mathcal{D}$.

Proof. (i) The assertion follows from $A \subset M$, compare the proof of Lemma 2.38(i).

(ii) We assume that $\bar{y}$ is a minimal element of A but not of M w.r.t. $\mathcal{D}$, i.e. there exist $y \in A$ and $d_y \in \mathcal{D}(y) \setminus \{0_Y\}$ with $y + d_y \in \{\bar{y}\} - (\mathcal{D}(\bar{y}) \setminus \{0_Y\})$. As $\mathcal{D}(\bar{y})$ is a pointed convex cone, this implies

$$y \in \{\bar{y}\} - (\mathcal{D}(y) \setminus \{0_Y\}) - (\mathcal{D}(\bar{y}) \setminus \{0_Y\})$$
$$\subset \{\bar{y}\} - (\mathcal{D}(\bar{y}) \setminus \{0_Y\}) - (\mathcal{D}(\bar{y}) \setminus \{0_Y\})$$
$$\subset \{\bar{y}\} - (\mathcal{D}(\bar{y}) \setminus \{0_Y\}),$$

in contradiction to $\bar{y}$ being a minimal element of A w.r.t. $\mathcal{D}$.

$\square$

The transitivity of the binary relation $\leq_2$ defined in (1.2) by the ordering map $\mathcal{D}$ is not sufficient for any minimal element of A being also a minimal element of M w.r.t. $\mathcal{D}$, in contrast to the result for nondominated elements. This is illustrated in the following example.

Example 2.49. Let Y be the Euclidean space $\mathbb{R}^2$, $A = \{(1,0),(1,1)\}$ and let the ordering map $\mathcal{D}: \mathbb{R}^2 \to 2^{\mathbb{R}^2}$ be defined by

$$\mathcal{D}(y_1, y_2) := \begin{cases} \text{cone}\{(-1,1)\} & \text{if } (y_1, y_2) \in \{(1,0) + \lambda(1,-1) \mid \lambda \geq 0\}, \\ \text{cone}\{(1,1)\} & \text{if } (y_1, y_2) \in \{(1,1) + \lambda(1,1) \mid \lambda \in \mathbb{R}\}, \\ \{0_Y\} & \text{otherwise.} \end{cases}$$

Then it is easy to verify that the relation $\leq_2$ is transitive according to Lemma 1.10(iii). The point $(1,1)$ is a minimal element of A w.r.t. $\mathcal{D}$, but the set M as defined in Lemma 2.48 equals

$$M = \{(1,0) + \lambda(-1,1) \mid \lambda \geq 0\} \cup \{(1,1) + \lambda(1,1) \mid \lambda \geq 0\},$$

and thus $(1/2, 1/2) \in M \cap (\{(1,1)\} - \mathcal{D}((1,1)))$. So $(1,1)$ is not a minimal element of M w.r.t. $\mathcal{D}$.

Note that in Lemma 2.48(i) above we have assumed that the minimal element $\bar{y}$ of M is also an element of A. Under additional assumptions we can show that any minimal element $\bar{y}$ of M is always an element of the set A.

Lemma 2.50. *Define M as in Lemma 2.48. If $\bar{y} \in M$ is a minimal element of M w.r.t. $\mathcal{D}$ and if*

$$d \in \mathcal{D}(y + d) \text{ for all } y \in A \text{ and for all } d \in \mathcal{D}(y), \tag{2.11}$$

then $\bar{y}$ is also a minimal element of A w.r.t. $\mathcal{D}$.

Proof. With Lemma 2.48 it remains to be shown that $\bar{y} \in A$. For this, we assume that $\bar{y} \in M \setminus A$, i.e. there exist $\bar{a} \in A$ and $\bar{d}_a \in \mathcal{D}(\bar{a}) \setminus \{0_Y\}$ with $\bar{y} = \bar{a} + \bar{d}_a$. Then $\bar{d}_a \in \mathcal{D}(\bar{y})$ and for $y := \bar{a} + \frac{1}{2}\bar{d}_a \in M \setminus \{\bar{y}\}$ we obtain

$$y = \bar{y} - \frac{1}{2}\bar{d}_a \in \{\bar{y}\} - (\mathcal{D}(\bar{y}) \setminus \{0_Y\}),$$

in contradiction to the minimality of $\bar{y}$ for the set M w.r.t. $\mathcal{D}$.

$\square$

The condition (2.11) is satisfied if

$$\mathcal{D}(y + d) \supset \mathcal{D}(y) \text{ for all } y \in A \text{ and for all } d \in \mathcal{D}(y). \qquad (2.12)$$

The ordering map presented in Example 1.12 satisfies also condition (2.12) and thus condition (2.11). We reconsider Example 2.10 where the ordering map $\mathcal{D}$ does not satisfy (2.11).

Example 2.51. Let Y, A and $\mathcal{D}$ be specified as in Example 2.10. For this set and ordering map we have already presented the set M, which is illustrated in Fig. 2.7, in Example 2.39. The set of minimal elements of M w.r.t. $\mathcal{D}$ is

$$\{(y_1, y_2) \in \mathbb{R}^2 \mid y_1 \in [-1, 0), y_2 = -\sqrt{1 - y_1^2}\}$$
$$\cup \{(y_1, y_2) \in \mathbb{R}^2 \mid y_1 > 0, y_2 = -1 - y_1\}.$$

Hence the set of minimal elements of M w.r.t. $\mathcal{D}$ intersected with A equals the set of minimal elements of A w.r.t. $\mathcal{D}$. But for instance $(1, -2)$ is a minimal element of M w.r.t. $\mathcal{D}$ but not of A.

The condition (2.11) does not hold: for $y = (0, -1)$, $d = (1, -1) \in \mathcal{D}(y)$ we obtain

$$d \notin \mathcal{D}(y + d) = \mathcal{D}((1, -2)) = \mathbb{R}_+^2.$$

Analogously to Lemma 2.42, we can relate minimal elements of a section of A w.r.t. $\mathcal{D}$ to minimal element of the set A w.r.t. $\mathcal{D}$.

Lemma 2.52. *Let $K \subset Y$ be a convex cone and consider for some $y \in Y$ the set $A_y = (\{y\} - K) \cap A$.*

(i) *If $\bar{y}$ is a minimal element of A_y w.r.t. $\mathcal{D}$ and $\mathcal{D}(\bar{y}) \subset K$, then $\bar{y}$ is also a minimal element of A w.r.t. $\mathcal{D}$.*

(ii) *Let $cor(\mathcal{D}(y)) \neq \emptyset$ for all $y \in A$. If $\bar{y}$ is a weakly minimal element of A_y w.r.t. $\mathcal{D}$ and $\mathcal{D}(\bar{y}) \subset K$, then $\bar{y}$ is also a weakly minimal element of A w.r.t. $\mathcal{D}$.*

Proof. (i) $\bar{y}$ minimal for A_y w.r.t. $\mathcal{D}$ is equivalent to $(\{\bar{y}\} - \mathcal{D}(\bar{y})) \cap A_y = \{\bar{y}\}$, thus to

$$(\{\bar{y}\} - \mathcal{D}(\bar{y})) \cap A \cap (\{y\} - K) = \{\bar{y}\}.$$

As $\{\bar{y}\} - \mathcal{D}(\bar{y}) \subset \{y\} - K - \mathcal{D}(\bar{y}) \subset \{y\} - K$ this implies $(\{\bar{y}\} - \mathcal{D}(\bar{y})) \cap A = \{\bar{y}\}$.

(ii) $\bar{y}$ weakly minimal for A_y w.r.t. $\mathcal{D}$ is equivalent to $(\{\bar{y}\} - cor(\mathcal{D}(\bar{y}))) \cap A_y = \emptyset$, thus to

$$(\{\bar{y}\} - cor(\mathcal{D}(\bar{y}))) \cap A \cap (\{y\} - K) = \emptyset.$$

As $\{\bar{y}\} - \text{cor}(\mathcal{D}(\bar{y})) \subset \{y\} - K - \text{cor}(\mathcal{D}(\bar{y})) \subset \{y\} - K$ this implies $(\{\bar{y}\} - \text{cor}(\mathcal{D}(\bar{y}))) \cap A = \emptyset$. □

Choosing $K = \mathcal{D}(A)$ (assuming $\mathcal{D}(A)$ to be convex) in the above lemma, then any (weakly) minimal element of the section w.r.t. $\mathcal{D}$ is also a (weakly) minimal element of A w.r.t. $\mathcal{D}$.

We conclude this section by a result about the minimal elements of the sum of two sets.

Lemma 2.53. *Let $A,\ B \subset Y$ be nonempty subsets and let $\mathcal{D}$ satisfy (2.10). If $\bar{y} = \bar{y}^A + \bar{y}^B \in A + B$ with $\bar{y}^A \in A$, $\bar{y}^B \in B$ is a minimal element of $A + B$ w.r.t. the ordering map $\mathcal{D}$, then $\bar{y}^A$ is a minimal element of A w.r.t. the ordering map $\mathcal{D}$ and $\bar{y}^B$ is a minimal element of B w.r.t. the ordering map $\mathcal{D}$.*

Proof. We assume $\bar{y}^A$ is not a minimal element of A w.r.t. the ordering map $\mathcal{D}$. Then there exists $y \in A$ with $\bar{y}^A \in \{y\} + (\mathcal{D}(\bar{y}^A) \setminus \{0_Y\})$ and thus

$$\bar{y} = \bar{y}^A + \bar{y}^B \in \{y + \bar{y}^B\} + (\mathcal{D}(\bar{y}^A) \setminus \{0_Y\}).$$

With (2.10) we conclude $\mathcal{D}(\bar{y}^A) \subset \mathcal{D}(\bar{y})$ and thus

$$\bar{y} \in \underbrace{\{y + \bar{y}^B\}}_{\in A+B} + (\mathcal{D}(\bar{y}) \setminus \{0_Y\})$$

in contradiction to $\bar{y}$ a minimal element of $A + B$ w.r.t. the ordering map $\mathcal{D}$. The same for $\bar{y}^B$. □

For a variable ordering structure satisfying (2.10) see Example 2.45. The converse statement of Lemma 2.53 does in general not hold even not in the case of a partially ordered space [134, Remark 3.1.3].

2.5 Notes on the Literature

The definition of a nondominated element was first given by Yu in 1974 [158], see also the work [19] by him and his colleagues Bergstresser and Charnes from 1976 and his book [159]. It is also shortly mentioned in the book by Sawaragi et al. [134], by Chew in [32] and by Weidner in [149] as well as in a more recent survey about advances in preference modeling by Wiecek [154]. Note that the original definition by Yu is incorrectly cited in [30, p. 98], [29, Definition 1.11]. There, (2.5) is replaced by $A \cap (\{\bar{y}\} - \mathcal{D}(y)) = \{\bar{y}\}$ for all $y \in A$. In the definition of the nondominated elements, the cone

$$\mathcal{D}(y) = \{d \in Y \mid y + d \text{ is dominated by } y\} \cup \{0_Y\},$$

is also called domination cone or the set of dominated directions or domination factors for each element $y \in Y$, cf. [19, 154, 159]. In the definition of the minimal elements, the cone

$$\mathcal{D}(y) = \{d \in Y \mid y - d \text{ is preferred to } y\} \cup \{0_Y\}$$

can be interpreted as the cone of preferred directions, compare page 10.

The concept of minimal elements, denoted there as nondominated-like elements, was introduced by Chen in 1992 [28], see also [29, 30]. This notion is also used by Huang et al. [87] and by Li and Li [109]. The minimal elements are denoted as nondominated elements w.r.t. $\mathcal{D}$ by Engau in [57, 58] and as efficient elements by Xiao et al. in [155]. In addition to the Examples 2.9 and 2.10, further examples comparing the concepts of minimal and nondominated elements w.r.t. a variable ordering structure are given by Chen [30], Eichfelder and Ha [52] and Gebhardt [61].

The definition of an efficient element in a partially ordered space can be found—under different names—in many books on vector optimization, see for instance the books by Jahn [94] or by Eichfelder [42], where these optimal elements are denoted as minimal elements, or the books by Göpfert and Nehse [70], Luc [113] and Sawaragi et al. [134], where the name efficient elements is used. For a survey about the different names used by different authors see the survey given by Ehrgott in [39, Table 2.4].

The concept of an efficient solution of a multiobjective optimization problem was probably first introduced in the applied sciences by Edgeworth in 1881 [38] and Pareto in 1906 [127]. Therefore, efficient elements are also called *Edgeworth-Pareto optimal points*. This naming was first suggested by Stadler [137]. For a brief historical sketch of the early works of Edgeworth and Pareto we refer to the survey in [53]. Condition (2.2) is also used to define optimal elements for an arbitrary set K with $0_Y \in K$, see Bergstresser et al. [19] and Weidner [153], or even with $0_Y \notin K$, see Weidner [150]. For a literature survey about the usage of the concept given in (2.2) for K as a set, a cone or the nonnegative orthant we refer to Engau [58].

The definitions of weaker and stronger optimality notions in partially ordered spaces can also be found in most introductory books on vector optimization. Ha provides in [77, Definition 21.3] an extensive collection of the different weaker and stronger concepts known in the literature. A survey about the various notions of proper efficiency and results on the contingent cone are given for instance in [94]. The definition of proper efficiency according to Henig is (of course) by Henig [83]. The definition according to Benson is from [16] while the definition according to Borwein is from [23]. For instance in [134] several examples are provided for showing the differences between the various proper optimality notions. For studies on the relations between the different notions of proper optimality in a partially ordered space we refer to [75, 111].

The concepts of weakly nondominated, weakly minimal and strongly minimal elements, some of the provided examples as well as some of the characterizations were first provided in [46], for some results see also [47] and [49]. The notion of strongly nondominated elements was introduced in [52]. The various discussed concepts of properly minimal and properly nondominated elements were given in [54]. There, one can also find some basic results on relations between the different notions. A more detailed study is provided in [55].

In [140, 141], Soleimani and Tammer extend notions of approximate solutions from vector optimization with a partially ordered linear space to problems with variable ordering structures. Thereby, they do not assume the sets $\mathcal{D}(y)$ to be convex or cones but closed sets containing the zero element. The definitions given here can also be used under these assumptions

In [155], Xiao, Xiao and Liu define strongly minimal solutions, denoted there as strong efficient solutions, of a vector optimization problem with a vector-valued objective map $f\colon X \to Y$ in a different way than proposed here: they denote by that minimal solutions $\bar{x}$ with the additional property that there is no other feasible x with $f(\bar{x}) = f(x)$. Note that for instance in [155] the variable ordering structure is not defined on Y but on X by a cone-valued map $\mathcal{C}\colon X \to 2^Y$. Of course, a direct relation exists to the ordering map as presented here: If $\mathcal{D}$ is given, then we can set $\mathcal{C}(x) := \mathcal{D}(f(x))$ for all x.

For defining an optimal solution of the set-valued vector optimization problem we have chosen here the so-called vector approach, see for instance [94, Chap. 17] or [53, Sect. 1.4.1]. Also another approach exists, denoted set-approach, which is based on binary set-relations in the power set of the space. For instance, there is a binary relation known as set less or KNY order relation which has been independently introduced by Young [157] and Nishnianidze [124] and has been presented by Kuroiwa [106] in a slightly modified form. Recently, several new binary relations for defining solutions to set-valued optimization problems have been proposed by Jahn and Ha [95].

Lemma 2.17 was already given by Yu in [158, Lemma 5.1(iv)]. Results on the set $A + K$ for some set $A \subset Y$ and an ordering cone $K \subset Y$ in a partially ordered linear space Y as well as on sections of sets are collected for instance in the book by Jahn [94]. Examinations for A convex-like, i.e. $A + K$ convex, as well as for A satisfying other concepts of generalized convexity, are given for instance by Adán and Novo in [1,2]. If A is not a convex set but $A + K$ is, for instance necessary linear scalarization results can be formulated which cannot be derived directly for A. The notion of external stability is taken from the book by Sawaragi et al. [134]. This notion is also denoted domination property, cf. the book by Luc [113] and the references therein. In this book and also in [134] several sufficient conditions ensuring the external stability of a set of efficient or nondominated elements are given. For the special case $Y = \mathbb{R}^m$, $\mathbb{R}^m_+ \subset K$ and compact sets A, the results of Theorems 2.20 and 2.22 were already given by Hirsch et al. in [86]. There, these results were used to solve vector optimization problems with a variable ordering structure by applying multiobjective evolutionary algorithms, cf. Sect. 9.3. The results in the way they are given here are due to [49].

Chapter 3
Cone-Valued Maps

Variable ordering structures are represented by cone-valued maps, i.e. by set-valued maps where the images are cones. For studying vector optimization problems with a variable ordering structure, it is important to have information about the properties of these cone-valued maps.

Thus, in the following we discuss for several well-known properties introduced in the literature for set-valued maps whether they are also applicable for cone-valued maps, i.e. whether assuming them is a too strong or a too weak assumption. For instance, convexity of a cone-valued map implies that the map is constant, i.e. this assumption is too strong. In case of non-appropriateness of the notions we propose new, more applicable, concepts. By that we provide tools for describing cone-valued maps. This is, for instance, used for defining criteria which guarantee that scalarization functionals which we discuss in the following chapters are convex. Again, this is important for formulating sufficient optimality conditions of Fermat and Lagrange type for vector optimization problems with such a variable ordering structure.

We especially pay attention to those cone-valued maps which have convex images, like we assume it in general for the ordering maps defining a variable ordering structure. Further, we mainly concentrate on those properties of cone-valued maps which are of interest for the study of the variable ordering structures.

In the following, when we consider set-valued maps F on a set $S \subset X$ or on the real linear space X we always assume $F(x) \subset Y$ to be nonempty for any $x \in S$ or $x \in X$, respectively. Here, Y is assumed to be a real linear space. $F(S)$ denotes the image of S under F, i.e. $F(S) = \bigcup_{x \in S} F(x)$.

3.1 Convexity and Linearity of Cone-Valued Maps

In this section, let X, Y be real linear spaces, if not stated otherwise, with S a nonempty algebraically open convex subset of X and $C \subset Y$ a convex cone.

G. Eichfelder, *Variable Ordering Structures in Vector Optimization*, Vector Optimization,   57
DOI 10.1007/978-3-642-54283-1_3, © Springer-Verlag Berlin Heidelberg 2014

3.1.1   Cone-Convex Cone-Valued Maps

We start by recalling the definition of a C-convex set-valued map for C a convex cone.

Definition 3.1. A set-valued map $F: S \to 2^Y$ is called C-convex, if for all $x^1, x^2 \in S$ and $\lambda \in [0, 1]$

$$\lambda F(x^1) + (1 - \lambda) F(x^2) \subset F(\lambda x^1 + (1 - \lambda) x^2) + C. \qquad (3.1)$$

If F is $\{0_Y\}$-convex, then F is called convex.

The next lemma gives a first characterization.

Lemma 3.2. *If a set-valued map $F: S \to 2^Y$ is C-convex, then $F(x) + C$ is convex for all $x \in S$.*

Proof. For any $x \in S$ we obtain from (3.1), by choosing $x^1 = x^2 = x$ and by using that C is a convex cone, that

$$\lambda (F(x) + C) + (1 - \lambda)(F(x) + C) \subset F(x) + C \quad \text{for all } \lambda \in [0, 1].$$

□

Recall that we have assumed the set S to be algebraically open.

Lemma 3.3. *Let $F: S \to 2^Y$ be a cone-valued map. If F is C-convex, then*

$$F(x) + C = F(S) + C \qquad \text{for all } x \in S. \qquad (3.2)$$

Proof. It holds $\lambda F(x) = F(x)$ and $0_Y \in F(x)$ for all $x \in S$ and all $\lambda > 0$. Thus (3.1) implies

$$\begin{aligned} F(x^2) &\subset F(x^2) + \lambda F(x^1) - \lambda F(x^2) \\ &\subset F(\lambda x^1 + (1 - \lambda) x^2) + C \end{aligned} \qquad (3.3)$$

for all $x^1, x^2 \in S$, $\lambda \in (0, 1)$. Let $\bar{x} \in \text{cor}(S) = S$ be given and choose $x \in S$ arbitrarily. Then there is some $\mu > 0$ such that

$$x^1 := \bar{x} + \mu(\bar{x} - x) \in S.$$

For $x^2 := x$ and $\lambda := 1/(1 + \mu)$, the inclusion (3.3) implies

$$F(x) \subset F\left(\frac{1}{1 + \mu} (\bar{x} + \mu(\bar{x} - x)) + \frac{\mu}{1 + \mu} x \right) + C = F(\bar{x}) + C$$

and thus

$$F(S) \subset F(\bar{x}) + C.$$

As S is algebraically open, there is some $v > 0$ such that

$$x^1 := x + v(x - \bar{x}) \in S.$$

For $x^2 := \bar{x}$ and $\lambda := 1/(1 + v)$, the inclusion (3.3) implies

$$F(\bar{x}) \subset F\left(\frac{1}{1+v}(x + v(x - \bar{x})) + \frac{v}{1+v}\bar{x}\right) + C = F(x) + C,$$

i.e. $F(S) + C \subset F(\bar{x}) + C \subset F(x) + C \subset F(S) + C$ and the assertion is proven.
□

The following theorem states that depending on the cone C the C-convexity of a cone-valued map might be a too strong assumption.

Theorem 3.4. *Let $F: S \to 2^Y$ be a cone-valued map with $F(x)$ a convex cone for all $x \in S$ and let*

$$C \subset \bigcap_{x \in S} F(x).$$

If F is C-convex, then F is a constant cone-valued map, i.e. $F(x) = K$ for all $x \in S$ and for some convex cone $K \subset Y$.

Proof. Let $x \in S$ be arbitrarily chosen. Because of the assumption on C and since $F(x)$ is a convex cone and $0_Y \in C$, we conclude

$$F(x) + C \subset F(x) + F(x) \subset F(x) \subset F(x) + C,$$

i.e. $F(x) = F(x) + C$. Using the result of Lemma 3.3, this implies

$$F(x) = F(x) + C = F(S) + C = \bigcup_{x \in S}(F(x) + C) = \bigcup_{x \in S} F(x) = F(S)$$

and the assertion is proven for $K = F(S)$.
□

Corollary 3.5. *Let $F: S \to 2^Y$ be a cone-valued map. If F is convex, then F is a constant cone-valued map, i.e. $F(x) = K$ for all $x \in S$ and for some convex cone $K \subset Y$.*

Proof. By Lemma 3.2, as F is convex, i.e. F is $\{0_Y\}$-convex, $F(x)$ is a convex cone for all $x \in S$ and with Theorem 3.4 the result follows.
□

Thus, assuming convexity of a cone-valued map implies that the map is constant. In Chap. 5 about nonlinear scalarization functionals we will see that the convexity of the ordering map of the variable ordering structure would guarantee convexity of some scalarization functionals—what can hence not be assumed.

On the other hand, for $F(x) \subset C$ for all $x \in S$, F is always C-convex, i.e. C-convexity is a redundant assumption.

Lemma 3.6. *Let $F \colon S \to 2^Y$ be a cone-valued map and let $F(x) \subset C$ for all $x \in S$. Then F is C-convex.*

Proof. For any $x^1, x^2 \in S$, $\lambda \in (0, 1)$ it holds

$$\lambda\, F(x^1) + (1 - \lambda)\, F(x^2) = F(x^1) + F(x^2) \subset C + C \subset C \subset \{0_Y\} + C$$
$$\subset F(\lambda x^1 + (1 - \lambda)x^2) + C.$$

$\square$

For $C \not\subset F(x)$ for some $x \in S$ and also $F(x) \not\subset C$ for some $x \in S$, a C-convex map is given in the following example:

Example 3.7 ([130]). Consider the cone-valued map $F \colon [0, \pi/2] \to 2^{\mathbb{R}^2}$,

$$F(x) = \{(r \cos \varphi, r \sin \varphi) \in \mathbb{R}^2 \mid r \ge 0,\ \varphi \in [0, x]\}$$

on $S = [0, \pi/2]$. For $C = \{(0, y) \in \mathbb{R}^2 \mid y \ge 0\}$ we obtain $F(x) + C = \mathbb{R}_+^2$ for all $x \in S$. As $F(x) \subset \mathbb{R}_+^2$ for all $x \in S$, (3.1) is satisfied for all $x^1, x^2 \in S, \lambda \in [0, 1]$ and F is C-convex.

The following lemma states that a modification of (3.1) by a restriction on bounded subsets of the cones—the cones intersected with the unit ball—, instead of the unbounded cones, delivers still not an appropriate concept for C-convexity of a cone-valued map F with convex cones as image sets and $C \subset F(x)$ for all $x \in S$.

Lemma 3.8. *Let $(Y, \|\cdot\|_Y)$ be a real normed space and let $F \colon S \to 2^Y$ be a cone-valued map with $F(x)$ a convex cone and $C \subset F(x)$ for all $x \in S$. If F satisfies for all $x^1, x^2 \in S$ and $\lambda \in [0, 1]$*

$$\lambda\left(F(x^1) \cap B_Y\right) + (1 - \lambda)\left(F(x^2) \cap B_Y\right) \subset \left(F(\lambda x^1 + (1 - \lambda)x^2) \cap B_Y\right) + C \tag{3.4}$$

with $B_Y \subset Y$ the closed unit ball, then $F(x) = K$ for all $x \in S$ and for some convex cone $K \subset Y$.

Proof. As $0_Y \in F(x) \cap B_Y$ for any $x \in S$, the inclusion (3.4) and the assumption on C imply for all $x^1, x^2 \in S, \lambda \in (0, 1)$:

$$(1 - \lambda)(F(x^2) \cap B_Y) \subset \left(F(\lambda x^1 + (1 - \lambda)x^2) \cap B_Y\right) + C$$
$$\subset F(\lambda x^1 + (1 - \lambda)x^2) + F(\lambda x^1 + (1 - \lambda)x^2)$$
$$\subset F(\lambda x^1 + (1 - \lambda)x^2).$$

Considering the cone generated by the set $(1 - \lambda)(F(x^2) \cap B_Y)$ yields

$$F(x^2) \subset F(\lambda x^1 + (1 - \lambda)x^2).$$

With the same arguments as in the proof of Lemma 3.3 we obtain for an arbitrary $\bar{x} \in S$ that $F(\bar{x}) = F(S) =: K$. □

For overcoming the drawbacks shown in Theorem 3.4, Lemmas 3.6 and 3.8, we modify the definition of C-convexity of a cone-valued map F as follows: instead of considering the cones $F(x)$ we give a definition via the bases of the cones $F(x)$, to be more concrete, via the vector-valued map which associates with each $x \in S$ the linear functional ℓ_x defining a base of F. For the definition of a base see page 2. For simplicity we assume Y to be a topological space and that all cones $F(x)$ for $x \in S$ have a base. Any cone having a base is pointed and a subset B of K is a base of K if and only if there is a continuous linear functional

$$\ell \in K^{\#} = \{y^* \in Y^* \mid y^*(y) > 0 \text{ for all } y \in K \setminus \{0_Y\}\}$$

with

$$B = \{y \in K \mid \ell(y) = 1\}.$$

According to the Krein-Rutman-Theorem (Theorem 1.20), every nontrivial closed pointed convex cone in a real separable normed space has a base. Recall also that a vector-valued map $f: S \to Y$ is C-convex for some convex cone $C \subset Y$ if

$$\lambda f(x^1) + (1 - \lambda)f(x^2) \in \{f(\lambda x^1 + (1 - \lambda)x^2)\} + C \ \forall \ x^1, x^2 \in S, \ \lambda \in [0, 1]. \tag{3.5}$$

Definition 3.9. Let Y be a real topological linear space, S a nonempty convex subset of X and $F: S \to 2^Y$ a cone-valued map with $F(x)$ a pointed convex cone having a base for each $x \in S$. If there is a C^*-convex map $\ell: S \to Y^*$ with the property that

$$B(x) := \{y \in F(x) \mid \ell(x)(y) = 1\} \tag{3.6}$$

is a base of $F(x)$ for all $x \in S$, then F is called C-baseconvex.

Example 3.10. Let X be a real linear space with S a nonempty convex subset and let $(Y, \| \cdot \|_Y)$ be a real normed space. Let $\ell: S \to Y^*$ be an arbitrary C^*-convex map with $\|\ell(x)\|_* > 1$ for all $x \in S$. $\| \cdot \|_*$ denotes the dual norm of $\| \cdot \|_Y$. We consider the cone-valued map $F: S \to 2^Y$ with images as Bishop-Phelps cones, cf. Sect. 1.2.1, defined by

$$F(x) = \mathcal{C}(\ell(x)) = \{y \in Y \mid \|y\|_Y \le \ell(x)(y)\} \text{ for all } x \in S.$$

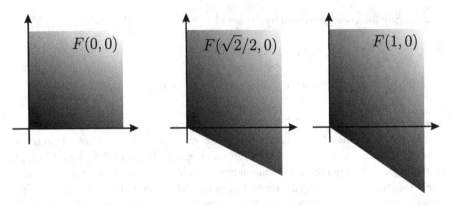

Fig. 3.1 The cones $F(0,0)$, $F(\sqrt{2}/2,0)$ and $F(1,0)$ of Example 3.10, cf. [48]

The sets $F(x)$ are nontrivial closed pointed convex cones and according to Lemma 1.16 $\ell(x)$ defines by (3.6) a base for $F(x)$ for all $x \in S$ and thus F is C-baseconvex on S.

For instance, let $S = X = Y = \mathbb{R}^2$ be Euclidean spaces with $C = \mathbb{R}^2_+ = C^*$. Define the map $\ell \colon \mathbb{R}^2 \to \mathbb{R}^2$ by

$$\ell(x_1, x_2) = (1 + x_1^2, 1 + x_2^2)^\top \quad \text{for all } (x_1, x_2) \in \mathbb{R}^2.$$

Then the cones $F(x) = \mathcal{C}(\ell(x))$ have a base defined by $\ell(x)$ and ℓ is C^*-convex and thus F is C-baseconvex on $\mathbb{R}^2$. As $\|\ell(x)\|_2 > 1$ for all $x \in \mathbb{R}^2$, the cones $F(x)$ have a nonempty interior according to Lemma 1.16. See Fig. 3.1 for an illustration of some of the cones. Additionally, $C = \mathbb{R}^2_+ \subset F(x)$ for all $x \in \mathbb{R}^2$. As $F(x) \neq K$ for some cone K, according to Theorem 3.4, F is not C-convex.

We end this subsection by examining cone-valued maps which satisfy a condition which is closely related to the convexity of the map. We need this characterization for the study of the convexity of the nonlinear scalarization functional introduced in Sect. 5.1.1.

Lemma 3.11. *Let* $F \colon X \to 2^Y$ *be a cone-valued map with*

$$F(x^i) \subset F(\lambda\, x^1 + (1 - \lambda)\, x^2), \qquad i = 1, 2 \tag{3.7}$$

for all $x^1, x^2 \in X$, $\lambda \in (0, 1)$. *Then* $F(x) = K$ *for all* $x \in X$ *and some cone* $K \subset Y$.

Proof. Let $x \in X$ be arbitrarily given. With $x^1 = x$, $x^2 = -x$ and $\lambda = 1/2$, (3.7) implies

$$F(x) \subset F(0_X)$$

and with $x^1 = 2x$, $x^2 = 0_X$ and $\lambda = 1/2$, (3.7) implies

$$F(0_X) \subset F(x)$$

and hence $F(x) = F(0_X) =: K$ for all $x \in X$. □

3.1.2 Convex-Like and Quasiconvex Cone-Valued Maps

As we have seen, cone-convexity of a cone-valued map is in many cases a too strong concept. Therefore, we consider in the following weaker concepts as convex-like maps. Again, let $C \subset Y$ be a convex cone.

Definition 3.12. A set-valued map $F: S \rightarrow 2^Y$ is called *convex-like* w.r.t. C if

$$\lambda F(x^1) + (1 - \lambda) F(x^2) \subset F(S) + C \; \forall \; x^1, x^2 \in S, \; \lambda \in (0, 1).$$

F is convex-like w.r.t. C if and only if $F(S) + C$ is a convex set [105, Definition 2.1], which is not a strong assumption if we assume F to be a cone-valued map with $F(x)$ a convex cone for all $x \in S$. In the next example a convex-like cone-valued map is given:

Example 3.13. For the cone-valued map $F: \mathbb{R}^2 \rightarrow 2^{\mathbb{R}^2}$ defined by

$$F(x_1, x_2) := \begin{cases} \left\{ \begin{pmatrix} r \cos \varphi \\ r \sin \varphi \end{pmatrix} \mid r \geq 0, \; \varphi \in [0, \frac{\pi}{4}] \right\} & \text{if } x_1 \geq \frac{\pi}{2}, \\ \left\{ \begin{pmatrix} r \cos \varphi \\ r \sin \varphi \end{pmatrix} \mid r \geq 0, \; \varphi \in [0, \frac{\pi}{2} + \frac{\pi}{4} - x_1] \right\} & \text{if } x_1 \in (\frac{\pi}{4}, \frac{\pi}{2}), \\ \mathbb{R}^2_+ & \text{if } x_1 \leq \frac{\pi}{4}, \end{cases}$$

see Fig. 3.2, it holds $F(\mathbb{R}^2) = \mathbb{R}^2_+$ and thus F is, for instance, convex-like w.r.t. any convex cone $C \subset \mathbb{R}^2_+$ or $\mathbb{R}^2_+ \subset C$.

Also the cone-quasiconvexity of a cone-valued map is an appropriate notion:

Definition 3.14. A set-valued map $F: S \rightarrow 2^Y$ is called C-*quasiconvex* if

$$(F(x^1) + C) \cap (F(x^2) + C) \subset F(\lambda x^1 + (1 - \lambda)x^2) + C$$

for all $x^1, x^2 \in S$, $\lambda \in (0, 1)$.

If F is C-convex then it is also C-quasiconvex [105, Proposition 2.1]. Thus, by Lemma 3.6, a cone-valued map F with $F(x) \subset C$ for all $x \in S$ is also C-quasiconvex. In contrast to $\{0_Y\}$-convex cone-valued maps also non-constant $\{0_Y\}$-quasiconvex maps exist:

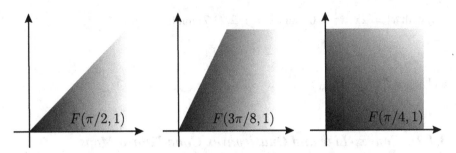

Fig. 3.2 The cones $F(\pi/2, 1)$, $F(3\pi/8, 1)$ and $F(\pi/4, 1)$ of Example 3.13, cf. [48]

Example 3.15. Consider F as specified in Example 3.13. For arbitrary $u = (u_1, u_2)$, $v = (v_1, v_2) \in \mathbb{R}^2$ with $u_1 \leq v_1$ it holds $F(v) \subset F(u)$. Thus, for arbitrary $u, v \in \mathbb{R}^2$ and arbitrary $\lambda \in (0, 1)$ we have w.l.o.g. $u_1 \leq v_1$ and then also $\lambda u_1 + (1 - \lambda)v_1 \leq v_1$ and we conclude

$$F(u) \cap F(v) = F(v) \subset F(\lambda u + (1 - \lambda)v)$$

and thus F is $\{0_Y\}$-quasiconvex.

This example shows that while $\{0_Y\}$-convexity of a cone-valued map is a too strong assumption (it implies that the cone-valued map is constant), non-constant quasiconvex maps exist. Hence, requiring quasiconvexity is not a too strong assumption.

3.1.3 Linear Cone-Valued Maps

We start with the definition of a linear set-valued map.

Definition 3.16. A set-valued map $F: X \to 2^Y$ with

$$\mu F(x^1) + \lambda F(x^2) = F(\mu x^1 + \lambda x^2) \quad \text{for all } x^1, x^2 \in X \text{ and } \mu, \lambda \in \mathbb{R} \quad (3.8)$$

is called *linear*.

Linearity is a stronger concept than convexity and according to Corollary 3.5 it is a too strong assumption for cone-valued maps. However note that linearity is already a strong assumption for set-valued maps. For instance, $F: X \to 2^Y$ linear implies $F(X)$ to be a subspace and $F(0_X) = \{0_Y\}$ and if F is constant and linear, then $F(x) \equiv \{0_Y\}$.

Lemma 3.17. *Let $F: X \to 2^Y$ be a linear cone-valued map. Then $F(x) = \{0_Y\}$ for all $x \in X$.*

Proof. Linearity in particular means that F is C-convex with $C = \{0_Y\}$, i.e. convex. Corollary 3.5 implies $F(x) = K$ for all $x \in X$ with K a convex cone. Then (3.8) reads as $\mu K + \lambda K = K$ for all $\mu, \lambda \in \mathbb{R}$ and $\mu = \lambda = 0$ imply $K = \{0_Y\}$. $\qquad\square$

Therefore, analogous to Definition 3.9, we modify the concept of linearity and adapt it to cone-valued maps by defining linearity via the linearity of a vector-valued map defining a base for each cone $F(x)$.

Definition 3.18. Let Y be a real topological linear space, S a nonempty subset of X and $F: S \to 2^Y$ a cone-valued map with $F(x)$ a pointed convex cone having a base for each $x \in S$. If there is a linear map $\ell: X \to Y^*$ with (3.6) is a base for $F(x)$ for all $x \in S$, then F is called *baselinear*.

We give an example for a baselinear cone-valued map.

Example 3.19. We consider again a cone-valued map F with images Bishop-Phelps cones, compare Example 3.10. Let $X = Y = \mathbb{R}^2$ be Euclidean spaces and $S = \{(x_1, x_2) \in \mathbb{R}^2 \mid x_1, x_2 \geq 1\}$. Define $\ell: \mathbb{R}^2 \to \mathbb{R}^2$ by $\ell(x) = Lx$ for all $x \in \mathbb{R}^2$ with

$$L := \begin{pmatrix} 4 & 2 \\ 2 & 3 \end{pmatrix}.$$

As ℓ is a linear map, $F: S \to 2^{\mathbb{R}^2}$ with

$$F(x) = \{y \in \mathbb{R}^2 \mid \|y\|_2 \leq (Lx)^\top y\} \text{ for all } x \in S$$

is baselinear.

However, baselinearity is still a strong assumption as it cannot be defined for any set S with $0_X \in S$: As linearity of ℓ implies $\ell(0_X) = 0_{Y^*}$ this would mean $B(x) = \emptyset$ in contradiction to $B(x)$ defining a base. We use the assumption of baselinearity in Lemma 6.7 to show the convexity of a scalarization functional.

3.1.4 Sublinear Cone-Valued Maps

In this subsection we study for a cone-valued map $F: X \to 2^Y$ the properties of subadditivity and positive homogeneity, which are related to the notion of a convex process. As before, let $C \subset Y$ be a convex cone.

Definition 3.20. A set-valued $F: X \to 2^Y$ with

$$F(x^1) + F(x^2) \subset F(x^1 + x^2) + C \text{ for all } x^1, x^2 \in X \tag{3.9}$$

and

$$F(\lambda x) = \lambda F(x) \text{ for all } \lambda > 0 \text{ and } x \in X \tag{3.10}$$

is called C-*sublinear*.

A set-valued map satisfying (3.10) together with $0_Y \in F(0_X)$ is called a *process*.

Definition 3.21. A $\{0_Y\}$-sublinear set-valued map $F: X \to 2^Y$ with $0_Y \in F(0_X)$ is called a *convex process*.

It is obvious that a C-sublinear set-valued map is C-convex. Also, one can easily verify that the C-convexity of a cone-valued map together with (3.10) implies (3.9). With Theorem 3.4 and Corollary 3.5 we conclude:

Corollary 3.22. *Let* $F: X \to 2^Y$ *be a* C-*sublinear cone-valued map.*

(a) *If* $F(x)$ *is a convex cone and* $C \subset F(x)$ *for all* $x \in X$, *then* $F(x) = K$ *for all* $x \in X$ *and for some convex cone* $K \subset Y$.

(b) *If* $C = \{0_Y\}$, *i.e.* F *is a convex process, then* $F(x) = K$ *for all* $x \in X$ *and for some convex cone* $K \subset Y$.

A cone-valued convex process F satisfies (3.9) with $C = \{0_Y\}$ and (3.10), which implies that F is convex and thus that F is a constant map. But already (3.9) with $C = \{0_Y\}$ implies this fact for arbitrary set-valued maps with $0_Y \in F(x)$ for all $x \in X$ and thus in particular for cone-valued maps:

Lemma 3.23. *Let* $F: X \to 2^Y$ *be a set-valued map with* $0_Y \in F(x)$ *for all* $x \in X$ *and let* F *satisfy (3.9) with* $C = \{0_Y\}$. *Then* $F(x) = \hat{F}$ *for all* $x \in X$ *and for some set* $\hat{F} \subset Y$.

Proof. For arbitrary $x \in X$ choose $x^1 := -x^2 := x$ in (3.9). Then $F(x) + F(-x) \subset F(0_X)$ and thus $F(x) \subset F(0_X)$ for all $x \in X$. We obtain

$$F(0_X) \subset F(X) =: \hat{F} \subset F(0_X).$$

This results in $F(0_X) = \hat{F}$. Setting $x^1 := x$, $x^2 := 0_X$, (3.9) leads to $F(x) + F(0_X) \subset F(x)$, i.e. $F(0_X) \subset F(x)$. Summarizing this we obtain $\hat{F} = F(0_X) \subset F(x) \subset \hat{F}$ for all $x \in X$.   □

For the cone C quite large in the sense that it contains all cones $F(x)$, C-sublinear cone-valued maps of course exist: if $F(x) \subset C$ for all $x \in X$, then (3.9) is always satisfied. And also non-constant cone-valued maps which satisfy (3.10) exist:

Example 3.24. We consider again the cone-valued map $\mathcal{D}: \mathbb{R}^2 \to 2^{\mathbb{R}^2}$ of Example 1.13. Then $\mathcal{D}(x) = \mathcal{D}(x/\|x\|_2)$ for all $x \in \mathbb{R}^2 \setminus \{0_{\mathbb{R}^2}\}$, thus $\mathcal{D}(x) = \mathcal{D}(\lambda x)$ for all $\lambda > 0$ and all $x \in X$ and this is equivalent to (3.10).

3.2 Continuity of Cone-Valued Maps

In this section we study the notion of semicontinuity for cone-valued maps as well as the local Lipschitz property. Let $(X, \| \cdot \|_X)$ and $(Y, \| \cdot \|_Y)$ be real normed spaces and let S be a nonempty open subset of X.

3.2.1 Semicontinuous Cone-Valued Maps

We start by recalling the definitions of upper and lower semicontinuous set-valued maps.

Definition 3.25. Let $F: S \to 2^Y$ be a set-valued map.

(a) F is called *lower semicontinuous at* $x^0 \in S$ if for all open sets $V \subset Y$ with $F(x^0) \cap V \neq \emptyset$ a neighborhood U of x^0 exists such that $F(x) \cap V \neq \emptyset$ for all $x \in U$. F is said to be *lower semicontinuous* if it is lower semicontinuous at any $x^0 \in S$.

(b) F is called *upper semicontinuous at* $x^0 \in S$ if for all open sets $V \subset Y$ with $F(x^0) \subset V$ a neighborhood U of x^0 exists such that $F(x) \subset V$ for all $x \in U$. F is said to be *upper semicontinuous* if it is upper semicontinuous at any $x^0 \in S$.

We give an example for an upper and a lower semicontinuous cone-valued map.

Example 3.26 ([130, Sect. 3.4.1]). Let $K_1 = \text{cone conv}\{(2, 1), (1, 2)\}$ and $K_2 = \mathbb{R}_+^2$. Define $F_1, F_2: \mathbb{R}^2 \to 2^{\mathbb{R}^2}$ by

$$F_1(x) = \begin{cases} K_1 & \text{if } x \in \mathbb{R}^2 \setminus \mathbb{Z}^2, \\ K_2 & \text{if } x \in \mathbb{Z}^2, \end{cases} \quad \text{and } F_2(x) = \begin{cases} K_2 & \text{if } x \in \mathbb{R}^2 \setminus \mathbb{Z}^2, \\ K_1 & \text{if } x \in \mathbb{Z}^2. \end{cases}$$

Then F_1 is upper but not lower semicontinuous and F_2 is lower but not upper semicontinuous.

In [22] an example was provided showing that upper semicontinuity is in general not an appropriate notion for cone-valued maps:

Example 3.27. Consider the cone-valued map $F: \mathbb{R}^n \to 2^{\mathbb{R}^2}$ defined by

$$F(x) = \begin{cases} K & \text{for all } x \in \mathbb{R}^n \setminus \{0_{\mathbb{R}^n}\}, \\ (-\infty, 0] \times \{0\} & \text{for } x = 0_{\mathbb{R}^n} \end{cases}$$

with K some cone in $\mathbb{R}^2$. For the open set

$$V = \{(y_1, y_2) \in \mathbb{R}^2 \mid y_1 < 1, \ |y_2| < \exp(y_1)\}$$

it holds $F(0_{\mathbb{R}^n}) \subset V$. If F is upper semicontinuous at $0_{\mathbb{R}^n}$, then there is a neighborhood U of $0_{\mathbb{R}^n}$ on which F can only take the values $F(0_{\mathbb{R}^n})$ or $\{0_{\mathbb{R}^2}\}$.

Thus, the upper semicontinuity of F implies $K = F(0_{\mathbb{R}^n})$, i.e. F to be constant, or $K = \{0_{\mathbb{R}^2}\}$, i.e. $F(x)$ to be trivial for all $x \in \mathbb{R}^n \setminus \{0_{\mathbb{R}^n}\}$.

We state this in a more general setting in the following lemma.

Lemma 3.28. *Let $(Y, \|\cdot\|_Y)$ be a real reflexive Banach space and $F: S \to 2^Y$ a cone-valued map with $F(x^0)$ a closed convex cone. F is upper semicontinuous at x^0 if and only if there is a neighborhood U of x^0 such that $F(x) \subset F(x^0)$ for all $x \in U$.*

Proof. The sufficiency of the condition is obvious. We show that the condition is also necessary. For that we assume there exists no neighborhood U of x^0 such that $F(x) \subset F(x^0)$ for all $x \in U$. Choose $\varepsilon > 0$ and consider the open neighborhood V_ε of $F(x^0)$ defined by

$$V_\varepsilon := \{z \in Y \mid \min_{y \in F(x^0)} \|z - y\|_Y < \varepsilon\}$$

(the above minimum over $F(x^0)$ always exists, cf. [92, Theorem 2.18]). Let U be an arbitrary neighborhood of x^0. Then some $x \in U \setminus \{x^0\}$ exists such that $F(x) \not\subset F(x^0)$. Thus some $y_U \in F(x)$ exists with $y_U \notin F(x^0)$. Set

$$\mu := \min_{y \in F(x^0)} \|y_U - y\|_Y > 0. \tag{3.11}$$

As $F(x)$ is a cone also $sy_U \in F(x)$ for all $s \geq \frac{\varepsilon}{\mu} > 0$. Then

$$\min_{y \in F(x^0)} \|sy_U - y\|_Y = s \min_{y \in F(x^0)} \|y_U - \frac{1}{s}y\|_Y = s\mu \geq \varepsilon,$$

i.e., $sy_U \notin V_\varepsilon$, and thus $F(x) \not\subset V_\varepsilon$ for some $x \in U$. As U was chosen arbitrarily this contradicts the upper semicontinuity of F at x^0. $\square$

We give an example showing that a result analogous to Lemma 3.28 does not hold for the lower semicontinuity of a cone-valued map.

Example 3.29. Let X, Y equal the Euclidean space $\mathbb{R}^2$ and let $S = \{(x_1, x_2) \in \mathbb{R}^2 \mid x_1 > 1,\ x_2 > 1\}$ and $F: S \to 2^{\mathbb{R}^2}$,

$$F(x_1, x_2) = \text{cone conv}\{(1,0), (x_1, x_2)\} \quad \text{for all } (x_1, x_2) \in S.$$

We can easily verify that F is lower semicontinuous on S: For that choose $x^0 \in S$ arbitrarily and an open set $V \subset \mathbb{R}^2$ with $F(x^0) \cap V \neq \emptyset$. Of interest are only sets V with $\partial F(x^0) \cap V \neq \emptyset$ (∂ denotes the boundary), otherwise the conclusion is obvious. Choosing $y^0 \in \partial F(x^0) \cap V$ arbitrarily there exists some $\delta > 0$ with

$$B_{y^0,\delta} := \{y \in \mathbb{R}^2 \mid \|y - y^0\|_2 < \delta\} \subset V.$$

As $y^0 \in \partial F(x^0)$, $y_2^0 = 0$ or $y^0 = sx^0$ for some $s > 0$. Only the second case is of interest. Then for

$$U := B_{x^0, \varepsilon} \cap S := \{x \in \mathbb{R}^2 \mid \|x - x^0\|_2 < \varepsilon\} \cap S$$

with $\varepsilon = \delta/s$ it holds for any $x \in U$ that $x \in F(x)$ by definition of F and thus also $sx \in F(x)$. As

$$\|sx - sx^0\|_2 = s\|x - x^0\|_2 < s\varepsilon = \delta$$

we obtain $sx \in B_{y^0, \delta} \subset V$, i.e. $sx \in F(x) \cap V$ and hence F is lower semicontinuous at x^0. For more details we refer to [130, Sect. 3.4.1].

For dealing with semicontinuity notions also for cone-valued maps F we define upper/lower semicontinuity of F via the map $F_B : S \to 2^Y$ defined by

$$F_B(x) = F(x) \cap B_Y \quad \text{for all } x \in S$$

(with B_Y the closed unit ball). A cone-valued map F is called *cosmically upper continuous* at $x^0 \in S$ if the map F_B is upper semicontinuous at x^0 [114]. By Proposition 2.1 in [114], a cone-valued map F is cosmically upper continuous at $x^0 \in S$ if and only if for every bounded closed set $B \subset Y$ the map $x \mapsto F(x) \cap B$ is upper semicontinuous at x^0. Further, also by Proposition 2.1 in [114], the map F_B is lower semicontinuous at some $x^0 \in S$ if and only if F is lower semicontinuous at x^0. The following example gives a cone-valued map F with F_B upper and lower semicontinuous:

Example 3.30. Let X, Y and F be specified as in Example 3.29 but with $S = (0, 1/2) \times (0, 1/2)$. Then the map F_B with

$$F_B(x_1, x_2) = \text{cone conv}\{(1, 0), (x_1, x_2)\} \cap B_Y \quad \text{for all } (x_1, x_2) \in S$$

is upper and lower semicontinuous on S. For more details we refer to [130, Sect. 3.4.1].

3.2.2 Lipschitz Continuous Cone-Valued Maps

We continue by recalling the definition of a locally Lipschitz map

Definition 3.31. A set-valued map $F : S \to 2^Y$ is called *locally Lipschitz* at $\bar{x} \in S$ with constant $\alpha > 0$ if there exists some neighborhood $U(\bar{x}) \subset S$ of $\bar{x}$ such that

$$F(x^1) \subset F(x^2) + \alpha \|x^1 - x^2\|_X B_Y \quad \text{for all } x^1, x^2 \in U(\bar{x}). \tag{3.12}$$

As before, B_Y denotes the closed unit ball in the real normed space Y.

Theorem 3.32. *Let $F: S \to 2^Y$ be a cone-valued map with $F(x)$ a closed cone for all $x \in S$. If F is locally Lipschitz at $\bar{x} \in S$ with constant $\alpha > 0$, then there is some neighborhood $U(\bar{x}) \subset S$ of $\bar{x}$ with $F(x) = K$ for all $x \in U(\bar{x})$ and with $K \subset Y$ a closed cone.*

Proof. Let $U(\bar{x}) \subset S$ and $\alpha > 0$ be given such that (3.12) is satisfied. Let $x^1, x^2 \in U(\bar{x})$, $x^1 \neq x^2$ be arbitrarily chosen and set

$$l := \alpha \|x^1 - x^2\|_X > 0.$$

Then

$$F(x^1) \subset F(x^2) + lB_Y. \tag{3.13}$$

Assuming $F(x^1) \neq F(x^2)$ there exists w.l.o.g. $\bar{y} \in F(x^1)$ with $\bar{y} \notin F(x^2)$. Then $\bar{y} \neq 0_Y$, as the images of F are cones. As $F(x^1)$ is a cone, also $\lambda \bar{y} \in F(x^1)$ for all $\lambda > 0$ and with (3.13) we get

$$\lambda \bar{y} \in F(x^2) + lB_Y \quad \text{for all } \lambda > 0.$$

This yields, because $F(x^2)$ is a cone, $\bar{y} \in F(x^2) + \frac{l}{\lambda}B_Y$. For λ to infinity this implies $\bar{y} \in \mathrm{cl}(F(x^2)) = F(x^2)$, which is a contradiction. $\square$

Therefore, the local Lipschitz property is generally not an appropriate concept for cone-valued maps. Instead, we define some kind of local Lipschitz property of a cone-valued map F via the map $F_B: S \to 2^Y$ defined by $F_B(x) = F(x) \cap B_Y$ for all $x \in S$. The following example gives a cone-valued map F with F_B locally Lipschitz continuous.

Example 3.33. Let X and Y be the Euclidean space $\mathbb{R}^2$ and consider the cone-valued map $F: \mathbb{R}^2 \to 2^{\mathbb{R}^2}$ defined by

$$F(x_1, x_2) = \{(y_1, y_2) \in \mathbb{R}^2 \mid y_1 = r \cos \varphi, \ y_2 = r \sin \varphi, \ r \geq 0,$$

$$\varphi \leq \tfrac{\pi}{2} + \min\{|x_2|, \tfrac{\pi}{6}\},$$

$$\varphi \geq -\min\{|x_1|, \tfrac{\pi}{6}\}\}.$$

Here, $\mathbb{R}^2_+ \subset F(x)$ for all $x \in \mathbb{R}^2$, see Fig. 3.3.

We show that the map F_B defined by $F_B(x) = F(x) \cap B_{\mathbb{R}^2}$ for all $x \in \mathbb{R}^2$ is locally Lipschitz on $\mathbb{R}^2$. For that we have to show that for any $\bar{x} \in \mathbb{R}^2$ there is some neighborhood $U(\bar{x})$ of $\bar{x}$ and some $\alpha > 0$ such that

$$F(x^1) \cap B_{\mathbb{R}^2} \subset F(x^2) \cap B_{\mathbb{R}^2} + \alpha \|x^1 - x^2\|_2 B_{\mathbb{R}^2} \quad \text{for all } x^1, x^2 \in U(\bar{x}).$$

Fig. 3.3 Images $F(x)$ for
$x \in (-\frac{\pi}{6}, \frac{\pi}{6}) \times (-\frac{\pi}{6}, \frac{\pi}{6})$ of
Example 3.33, cf. [48]

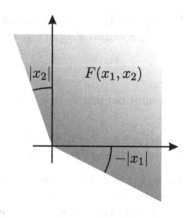

This can be done by straightforward calculations using that cosinus and sinus are
Lipschitz continuous functions and by using $\|x\|_2 \leq \|x\|_1$ for all $x \in \mathbb{R}^2$. For more
details we refer to [130, Sect. 3.4.2].

If $\bar{y}$ is given in $F(\bar{x})$ in Definition 3.31 then F is called *locally pseudo-Lipschitz*
around $(\bar{x}, \bar{y})$ with constant $\alpha \geq 0$ if there exists some neighborhood $U(\bar{x}) \subset S$ of
$\bar{x}$ and $V(\bar{y}) \subset Y$ of $\bar{y}$ such that

$$F(x^1) \cap V(\bar{y}) \subset F(x^2) + \alpha \|x^1 - x^2\|_X B_Y \quad \text{for all } x^1, x^2 \in U(\bar{x}). \tag{3.14}$$

This property is also known as *Aubin property*. A cone-valued map F can be locally
pseudo-Lipschitz without being locally constant:

Example 3.34. Let $F: S \to 2^Y$ be a cone-valued map with

$$\bar{y} \in \text{int}(\bigcap_{x \in S} F(x)).$$

Then F is locally pseudo-Lipschitz at any $(\bar{x}, \bar{y})$ with arbitrary constant $\alpha > 0$ and
arbitrary $\bar{x} \in X$.

Also, any cone-valued map F with F_B being locally Lipschitz continuous at $\bar{x}$,
is also locally pseudo-Lipschitz at $(\bar{x}, \bar{y})$ for any $\bar{y} \in \text{int} B_Y$.

3.3 Monotonicity of Cone-Valued Maps

In this section we discuss the notion of monotonicity for cone-valued maps. Let
$(Y, \langle \cdot, \cdot \rangle)$ be a real Hilbert space identified with its dual, if not stated otherwise, and
let S be an open nonempty subset of Y. Recall that a monotone set-valued map
$F: S \to 2^Y$ is defined by

Definition 3.35. A set-valued map $F: S \rightarrow 2^Y$ is called *monotone* if for all $y^1, y^2 \in S$ and all $u^1 \in F(y^1)$, $u^2 \in F(y^2)$

$$\langle u^1 - u^2, y^1 - y^2 \rangle \geq 0. \tag{3.15}$$

It turns out that all monotone cone-valued maps F are constant trivial-valued maps.

Theorem 3.36. *Let $F: S \rightarrow 2^Y$ be a monotone set-valued map with $0_Y \in F(y)$ for all $y \in S$. Then $F(y) = \{0_Y\}$ for all $y \in S$.*

Proof. Let $y \in S$ be arbitrarily chosen. As S is an open set, for any $h \in Y$ there exists some $\lambda > 0$ such that $y \pm \lambda h \in S$. Choose any $u \in F(y)$. First, set $y^1 := y + \lambda h$. Then according to (3.15) it holds for all $u^1 \in F(y^1)$

$$\langle u^1 - u, \lambda h \rangle \geq 0.$$

Choosing $u^1 := 0_Y \in F(y^1)$ we get $\langle u, h \rangle \leq 0$. Next, set $y^2 := y - \lambda h$. Then with (3.15) it holds for all $u^2 \in F(y^2)$,

$$\langle u - u^2, \lambda h \rangle \geq 0.$$

Again, as $0_Y \in F(y^2)$, we obtain $\langle u, h \rangle \geq 0$ and thus $\langle u, h \rangle = 0$ for all $h \in Y$ and hence $u = 0_Y$.   □

Corollary 3.37. *Let $F: S \rightarrow 2^Y$ be a monotone cone-valued map. Then $F(y) = \{0_Y\}$ for all $y \in S$.*

Analogously to C-baseconvex and baselinear, we define basemonotonicity. For that we need the following definition:

Definition 3.38. Let Y be a real topological linear space and S a nonempty subset of Y. We say that a map $f: S \rightarrow Y^*$ is *monotone* if

$$(f(y^1) - f(y^2))(y^1 - y^2) \geq 0 \text{ for all } y^1, y^2 \in S.$$

Definition 3.39. Let Y be a real topological linear space and $S \subset Y$ a nonempty subset. Let $F: S \rightarrow 2^Y$ be a cone-valued map with $F(y)$ a pointed convex cone having a base for any $y \in S$. If there is a monotone map $\ell: S \rightarrow Y^*$ with

$$B(y) = \{z \in F(y) \mid \ell(y)(z) = 1\}$$

is a base of $F(y)$ for any $y \in S$, then F is called *basemonotone*.

Example 3.40. We consider again the cone-valued map F with images being Bishop-Phelps cones as defined in Example 3.19. As L is a real symmetric positive semidefinite matrix, the map ℓ defined by $\ell(y) = Ly$ for all $y \in \mathbb{R}^2$ is monotone and thus F is also basemonotone.

Basemonotone and baselinear cone-valued maps play an important role as assumptions for guaranteeing the convexity of a scalarization functional, see Chap. 6.

3.4 Notes on the Literature

Cone-valued maps are not only important in view of variable ordering structures but play also an important role in optimization in general. For instance, optimality conditions based on contingent cones, see e.g. the book by Jahn [94, Theorem 3.48], associate with each $x \in S$, with S a nonempty subset of a real normed space, the contingent cone $T(S, x)$ to the set S at x. The operation taking the polar cone, also a cone-valued map, was studied by Walkup and Wets in [148]. In [22] the continuity of the cone-valued map associating with each x the normal cone to the related level set of a function f is discussed by Borde and Crouzeix. If f is convex and finite, then this cone-valued map equals the map associating with each x the cone generated by the subgradient of f at x. Also, the map associating with any sublinear functional its epigraph is a set-valued map with each image a convex cone. Further, the Bishop-Phelps cones as discussed in Sect. 1.2 describe a cone-valued map $C: Y^* \to 2^Y$: with any continuous linear functional ϕ on Y, with $(Y, \| \cdot \|_Y)$ a real normed space, the convex cone

$$C(\phi) = \{y \in Y \mid \|y\|_Y \le \phi(y)\},$$

denoted a Bishop-Phelps cone, is associated. In [94, p. 373] Jahn mentions a cone-valued map that describes the emission cone of an ultrasonic sensor of an autonomous transportation robot used for determining distances to obstacles.

The importance of the study of the properties of cone-valued maps becomes also obvious in vector optimization in a real reflexive Banach space Y partially ordered by a closed pointed convex cone $K \subset Y$. In [73] Guerra, Melguizo and Muñoz-Bouzo study the upper semicontinuity of the ideal conic map $C_A: S \to 2^Y$ which is related to the sensitivity of a vector optimization problem. Thereby, $A \subset Y$ is a given nonempty set and $S \subset Y$ denotes the set of all so called ideal points, i.e. those elements $y \in Y$ for which there exists a closed pointed convex cone $C_A(y) \subset Y$ with $K \subset C_A(y)$ and $A \subset \{y\} + C_A(y)$. The ideal conic map is then defined by

$$C_A(y) := \mathrm{cl}\,(\mathrm{conv}\,(\mathrm{cone}\,(A - \{y\}) \cup K)) \text{ for all } y \in S.$$

It was shown that the map is (if the space Y has a dimension larger than 1) not upper semicontinuous at the points $y \in S \cap A$ and not Lipschitz around any $y \in S \cap A$. A related map, the polar conic map $F_A: Y \to 2^{Y^*}$ given by

$$F_A(y) := \{y^* \in K^* \mid y^*(y) \le y^*(a) \text{ for all } a \in A\},$$

with $(F_A(y))^* = C_A(y)$ and $F_A(y) = (C_A(y))^*$ for all $y \in Y$ is examined by the same authors in [74] on upper and lower semicontinuity.

More related to variable ordering structures, cone-valued maps also appear in the context of vector complementarity problems, vector equilibrium problems and vector variational inequalities, see also Sect. 10.1. In [68], Giannessi, Mastroeni and Yang give a survey about vector complementarity problems with a variable ordering structure, recalling results of the book by Chen et al. [29]. Thereby they consider an ordering structure defined by a cone-valued map $F: X \rightarrow 2^Y$ with closed convex cones $F(x)$ as images which is assumed to be upper semicontinuous. Chen, Yang and Yu assume in [31] the upper semicontinuity of a cone-valued map for showing the upper semicontinuity of a scalarization functional in the context of vector equilibrium problems, see also Sect. 5.3. They provide a simple example for such a cone-valued map being constant everywhere but on a line (and constant on the line again), see [31, Example 2.2]. Zheng studies in [161] vector variational inequalities, see also Sect. 10.1, and for an existence result it is assumed that the set-valued map $W: S \rightarrow 2^Y$ is convex where $W(x) := Y \setminus -\text{int}(C(x))$ with $C(x)$ a closed pointed convex cone with nonempty interior for each $x \in S \subset X$ defining also a variable ordering structure in this context. W is thus also a cone-valued map.

The definition of a C-convex set-valued map and of a convex process as given here is taken from Aubin and Frankowska [8, Lemma 2.1.2], see also Kuroiwa [105, Definition 2.1] as well as Benoist and Popovici [15], and the references therein. We refer to Kuroiwa [105] for the study of several convexity notions for set-valued maps including convex-like and quasiconvex.

Linearity of a cone-valued map was presumed and defined in the way it is presented here by Chen and Yang in [30, p. 3] in the context of scalarization functionals for variable ordering structures. Assuming linearity of the cone-valued ordering map it could be shown that a nonlinear scalarization functional for vector optimization problems with variable ordering structures is convex. We present this scalarization functional, denoted also as Gerstewitz functional or translative functional, in Sect. 5.2. Chen et al. [29, Definition 1.70] also use the definition of a convex process as defined here for the study of vector optimization problems with a variable ordering structure. For a convex process we also refer to Lemma 2.1.2 in the book by Aubin and Frankowska [8].

Some definitions of properties of set-valued maps as local Lipschitz continuity, semicontinuity or monotonicity can be found in the book by Aubin and Frankowska [8, Sect. 1.4 and Definition 3.5.1]. The notion of upper and lower semicontinuity is thereby based on the one given by Berge [17]. For a survey on the different notions of upper and lower semicontinuity used in the literature and their relations we refer to the books by Aubin and Ekeland [7] and by Bank et al. [10] as well as to the survey by Delahaye and Denel [37].

Set-valued monotone maps as defined here are also of interest for auxiliary variational inequalities formulated to vector optimization problems in partially ordered spaces, compare page 277 in [27] by Ceng, Mordukhovich and Yao. There, a set-valued monotone map is defined by associating with each element the sum of the normal cone to some set at this element and another monotone map, which is then not cone-valued. Basemonotone and baselinear ordering maps play an important role for implying the convexity of a nonlinear scalarization functional, see Chap. 6.

The examinations for cone-valued maps as presented here are due to [48]. In [114], Luc and Penot have already shown that for any cone-valued map F defined on a metrizable topological space and being upper semicontinuous at some $x^0 \in S$, there exists a neighborhood U of x^0 such that $F(U) \subset \mathrm{cl}(F(x^0))$, which is a more general result than the one presented in Lemma 3.28. Examples of cone-valued maps satisfying the properties discussed in this chapter as well as an extensive study of the proposed concepts for cone-valued maps were given by Pruckner in his diploma thesis [130]. Example 3.27 is due to Borde and Crouzeix [22].

Similar considerations as summarized in this chapter are also done for set-valued maps with bounded images or with the property $F(0_X) = \{0_Y\}$ by Berge in [18] and by Godini in [69], see also the paper by Nikodem and Popa [123] and the references therein. It was shown, for instance, that such maps are single-valued under properties as convexity or linearity.

Chapter 4
Linear Scalarizations

Characterizing optimal elements of a vector optimization problem by scalarization, i.e. by the replacement of the vector optimization problem with a (parameter dependent) optimization problem with a scalar-valued objective function, is an important tool in vector optimization. By that, necessary and sufficient optimality conditions for single and set-valued vector optimization problems can be derived using the techniques from scalar-valued optimization. Also, existence and duality results and numerical solution approaches can be obtained using the characterization of optimal elements by scalarization functionals. We start with linear scalarizations before we discuss also nonlinear scalarization functionals in the next chapters.

Thus, this chapter is devoted to the characterization of nondominated and minimal elements of a set w.r.t. the ordering map $\mathcal{D}$ by means of linear scalarization functionals. Based on these characterizations we also formulate an existence result for nondominated and for minimal elements of a set w.r.t. the ordering map $\mathcal{D}$.

In the following we assume, as before, Y to be a real linear space equipped with a variable ordering structure defined by the ordering map $\mathcal{D}: Y \to 2^Y$ with $\mathcal{D}(y)$ a pointed convex cone. Additionally, let A be a nonempty subset of Y.

4.1 Characterization of Nondominated Elements

In this section we study necessary and sufficient conditions for (weakly) nondominated elements w.r.t. a variable ordering structure based on linear scalarization functionals. Because one has to check all sets $\{y\}+\mathcal{D}(y)$ to see whether they include the candidate element $\bar{y}$, strong assumptions have to be applied for a sufficient condition. Let

$$\mathcal{D}(A) = \bigcup_{y \in A} \mathcal{D}(y)$$

G. Eichfelder, *Variable Ordering Structures in Vector Optimization*, Vector Optimization, DOI 10.1007/978-3-642-54283-1_4, © Springer-Verlag Berlin Heidelberg 2014

denote the image of the set A under the set-valued map $\mathcal{D}$. Note that $\mathcal{D}(A)$ is a cone but not necessarily a convex cone. We formulate sufficient conditions with the help of elements of the dual cone

$$(\mathcal{D}(A))' = \{l \in Y' \mid l(y) \geq 0 \; \forall y \in \mathcal{D}(A)\}$$

of the cone $\mathcal{D}(A)$, which might reduce to the trivial cone if $\mathcal{D}(A)$ is not pointed. This is the case if the cones $\mathcal{D}(y)$ for $y \in A$ vary too much. In that case it will turn out that the sufficient conditions are too strong (and the necessary ones too weak) and linear scalarizations are not an adequate tool. Then the examination of other scalarizing functionals is important, which we will do in Chaps. 5 and 6. On the other hand, in many applications the cones $\mathcal{D}(y)$ may vary only slightly, as it is the case for the variable ordering structures examined in [58] or in the Examples 1.11 and 1.12, and then linear functionals may be useful.

We start with a necessary and a sufficient condition for weakly nondominated elements w.r.t. a variable ordering structure. In addition to the image set $\mathcal{D}(A)$ we need the convex cone

$$\hat{D} := \bigcap_{y \in A} \mathcal{D}(y).$$

For the necessary condition we need also a separation theorem:

Theorem 4.1 ([94, Theorem 3.14]). *Let S and T be nonempty convex subsets of a real linear space Y with $cor(S) \neq \emptyset$. Then $cor(S) \cap T = \emptyset$ if and only if there are a linear functional $l \in Y' \setminus \{0_{Y'}\}$ and a real number α with*

$$l(s) \leq \alpha \leq l(t) \text{ for all } s \in S \text{ and all } t \in T$$

and

$$l(s) < \alpha \text{ for all } s \in cor(S).$$

Theorem 4.2. *Let $cor(\mathcal{D}(y))$ be nonempty for all $y \in A$.*

(a) *Let A be convex as well as $cor(\hat{D}) \neq \emptyset$. Let $\bar{y}$ be a weakly nondominated element of A w.r.t. $\mathcal{D}$. Then a linear functional $l \in \hat{D}' \setminus \{0_{Y'}\}$ exists with $\bar{y}$ a minimal solution of $\min_{y \in A} l(y)$, i.e.*

$$l(\bar{y}) \leq l(y) \text{ for all } y \in A.$$

(b) *If for some $l \in (\mathcal{D}(A))' \setminus \{0_{Y'}\}$ the element $\bar{y} \in A$ is a minimal solution of $\min_{y \in A} l(y)$, i.e.*

$$l(\bar{y}) \leq l(y) \text{ for all } y \in A,$$

then $\bar{y}$ is a weakly nondominated element of A w.r.t. $\mathcal{D}$.

Proof. (a) Since $\bar{y} \in A$ is a weakly nondominated element of A w.r.t. $\mathcal{D}$, it holds that $\bar{y} \notin \{y\} + \text{cor}(\mathcal{D}(y))$ and thus $\bar{y} \notin \{y\} + \text{cor}(\hat{D})$ for all $y \in A$. Then $(\{\bar{y}\} - \text{cor}(\hat{D})) \cap A = \emptyset$ and with Theorem 4.1 this results in

$$l(y) \geq \alpha \geq l(\bar{y} - k) \text{ for all } y \in A, \ k \in \hat{D}$$

for some $\alpha \in \mathbb{R}$ and some $l \in Y' \setminus \{0_{Y'}\}$. As $0_Y \in \hat{D}$ we get $l(\bar{y}) \leq l(y)$ for all $y \in A$ and as $\bar{y} \in A$ also $l(-k) \leq 0$ for all $k \in \hat{D}$ and thus $l \in \hat{D}' \setminus \{0_{Y'}\}$.

(b) According to Lemma 1.7(i) and as $(\mathcal{D}(A))' \subset (\mathcal{D}(y))'$,

$$\text{cor}(\mathcal{D}(y)) \subset \{z \in Y \mid l(z) > 0 \text{ for all } l \in (\mathcal{D}(A))' \setminus \{0_{Y'}\}\}$$

for all $y \in A$. Thus it holds for any $l \in (\mathcal{D}(A))' \setminus \{0_{Y'}\}$ that $l(\bar{y}) > l(y)$ for all $y \in A$ with $\bar{y} - y \in \text{cor}(\mathcal{D}(y))$. Hence $l(\bar{y}) \leq l(y)$ implies $\bar{y} \notin \{y\} + \text{cor}(\mathcal{D}(y))$ for all $y \in A$.

$\square$

The following example provides a set A and an ordering map $\mathcal{D}$ where all elements satisfying the necessary optimality conditions are in fact weakly nondominated elements.

Example 4.3. Let Y, A and $\mathcal{D}$ be specified as in Example 2.10, i.e. $Y = \mathbb{R}^2$, $A := \{y \in \mathbb{R}^2 \mid \|y\|_2 \leq 1\}$ and

$$\mathcal{D}(y) := \begin{cases} \mathbb{R}_+^2 & \text{for } y \in \mathbb{R}^2 \setminus \{(0, -1)\}, \\ \{(z_1, z_2) \in \mathbb{R}^2 \mid z_1 + z_2 \geq 0, z_1 \geq 0\} & \text{for } y = (0, -1), \end{cases}$$

Then A is convex, $\hat{D} = \mathbb{R}_+^2$ has a nonempty algebraic interior and $\hat{D}' = \mathbb{R}_+^2$. We determine all elements $y \in A$ which satisfy the necessary condition of Theorem 4.2(a) for being a weakly nondominated element of A w.r.t. $\mathcal{D}$. Let $\bar{y} \in A$ be arbitrarily chosen. If there is some $y \in A$ with $y_i < \bar{y}_i$, $i = 1, 2$, then $\bar{y}$ cannot be a minimal solution of

$$\min_{y \in A} l_1 y_1 + l_2 y_2$$

for any $l \in \hat{D}' \setminus \{(0, 0)\} = \mathbb{R}_+^2 \setminus \{(0, 0)\}$. Thus, only elements of the set

$$\tilde{A} := \left\{ (y_1, y_2) \in \mathbb{R}^2 \,\middle|\, y_1 \in [-1, 0], \ y_2 = -\sqrt{1 - y_1^2} \right\}$$

are candidates for satisfying the necessary conditions.

Next, let $\bar{y} \in \tilde{A}$ be arbitrarily chosen. Then for

$$\bar{l} := \frac{1}{\|\bar{y}\|_2} \begin{pmatrix} |\bar{y}_1| \\ |\bar{y}_2| \end{pmatrix} = \begin{pmatrix} |\bar{y}_1| \\ |\bar{y}_2| \end{pmatrix} \in \mathbb{R}_+^2 \setminus \{(0, 0)\}$$

we obtain $\bar{l}^\top \bar{y} = -(\bar{y}_1^2 + \bar{y}_2^2) = -1$ and for any $y \in A$ we have

$$\bar{l}^\top y \geq -|\bar{l}^\top y| \geq -\|\bar{l}\|_2 \|y\|_2 = -\|y\|_2 \geq -1,$$

i.e. $\bar{y}$ is a minimal solution of $\min_{y \in A} \bar{l}^\top y$. Hence all elements $\bar{y} \in \tilde{A}$ satisfy the necessary condition discussed.

Here,

$$(\mathcal{D}(A))' = \{(y_1, y_2) \in \mathbb{R}_+^2 \mid y_2 \leq y_1\}.$$

For $(\bar{y}_1, \bar{y}_2) \in \tilde{A}$ with $\bar{y}_1 \leq \bar{y}_2$ and thus $\bar{y}_1 \leq \bar{y}_2 \leq 0$ we have for $\bar{l} = (|\bar{y}_1|, |\bar{y}_2|)^\top$ that $\bar{l} \in (\mathcal{D}(A))' \setminus \{(0,0)\}$ and $\bar{y}$ is a minimal solution of $\min_{y \in A} \bar{l}^\top y$. According to Theorem 4.2(b), $\bar{y}$ is a weakly nondominated element of A w.r.t. $\mathcal{D}$. The elements of the set $\{(y_1, y_2) \in \tilde{A} \mid y_1 > y_2\}$ are also weakly nondominated elements of A w.r.t. $\mathcal{D}$, see Example 2.10, but they do not satisfy the sufficient condition of Theorem 4.2(b).

However, the necessary condition for weakly nondominated elements w.r.t. the ordering map $\mathcal{D}$ may be very weak if the cones $\mathcal{D}(y)$ for $y \in A$ vary too much, because then the cone $\hat{D}$ is very small (or even trivial) and the dual cone is very large:

Example 4.4. Let $Y \in \mathbb{R}^2$ and let $\mathcal{D}$ and A be defined as in Example 2.35, i.e. $A = [1, 3] \times [1, 3]$ and

$$\mathcal{D}(y_1, y_2) := \begin{cases} \mathbb{R}_+^2 & \text{if } y_1 \geq 2, \\ \{(z_1, z_2) \in \mathbb{R}^2 \mid z_1 \leq 0, \ z_2 \geq 0\} & \text{otherwise.} \end{cases}$$

The unique nondominated element w.r.t. $\mathcal{D}$ is $(2, 1)$ and all the elements of the set

$$\{(2, t) \in \mathbb{R}^2 \mid t \in [1, 3]\} \cup \{(t, 1) \in \mathbb{R}^2 \mid t \in [1, 3]\}$$

are weakly nondominated w.r.t. $\mathcal{D}$. Further,

$$\mathcal{D}(A) = \{(z_1, z_2) \in \mathbb{R}^2 \mid z_2 \geq 0\}$$

and thus

$$(\mathcal{D}(A))' = \{(z_1, z_2) \in \mathbb{R}^2 \mid z_1 = 0, \ z_2 \geq 0\},$$

i.e. $(\mathcal{D}(A))^\# = \emptyset$.

Let $l \in (\mathcal{D}(A))' \setminus \{0_{Y'}\}$ be arbitrarily chosen, i.e. $l_1 = 0, l_2 > 0$, and consider the scalar-valued optimization problem $\min_{y \in A} l^\top y$. Then all elements of the set

$$\{(t, 1) \in \mathbb{R}^2 \mid t \in [1, 3]\}$$

Fig. 4.1 The set A of
Example 4.5, cf. [61]

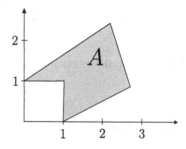

are minimal solutions and hence are weakly nondominated elements of A w.r.t. $\mathcal{D}$
according to Theorem 4.2(b). All the other weakly nondominated elements w.r.t. $\mathcal{D}$
cannot be found by the sufficient condition. Because of $\mathrm{int}(\hat{D}) = \emptyset$, the necessary
condition of Theorem 4.2(a) cannot be applied.

The next example, based on [61, Example 3.20], considers a nonconvex set A
where the necessary condition of Theorem 4.2(a) cannot be applied. It is a well
known fact that even for vector optimization problems in partially ordered spaces
for the necessary conditions based on linear scalarizations convexity assumptions
are required.

Example 4.5. Let $Y = \mathbb{R}^2$,

$$A := \{(y_1, y_2) \in \mathbb{R}^2 \mid (y_1 \geq 1 \vee y_2 \geq 1) \wedge -y_1 + y_2 \leq 1$$
$$\wedge \; y_1 - 2y_2 \leq 1 \wedge y_1 + y_2/3 \leq 3\},$$

i.e. $A \subset \mathbb{R}^2_+$, see Fig. 4.1, and

$$A^s := \{(y_1, y_2) \in \mathbb{R}^2 \mid y_1 - 2y_2 = 1, \; 1 \leq y_1 \leq 19/7\} \subset \partial A.$$

Since A is not convex, Theorem 4.2(a) cannot be applied. Let the ordering map
$\mathcal{D}: \mathbb{R}^2 \to 2^{\mathbb{R}^2}$ be defined by

$$\mathcal{D}(y) := \begin{cases} \{(z_1, z_2) \in \mathbb{R}^2 \mid z_2 \geq z_1 \geq 0\} & \text{for } y \in \mathbb{R}^2 \setminus A, \\ \{(z_1, z_2) \in \mathbb{R}^2 \mid z_1 \geq z_2 \geq 0\} & \text{for } y \in A \setminus \{(0, 1)\}, \\ \mathbb{R}^2_+ & \text{for } y = (0, 1). \end{cases}$$

In the following we only use that $\mathcal{D}(A) = \mathbb{R}^2_+$. We evaluate the sufficient condition
of Theorem 4.2(b). $\mathcal{D}(A) = \mathbb{R}^2_+$ and thus

$$(l_1, l_2) \in (\mathcal{D}(A))' \setminus \{0_{Y'}\} \;\Leftrightarrow\; l_1 \geq 0, \; l_2 \geq 0, \; (l_1, l_2) \neq (0, 0).$$

For $l \in (\mathcal{D}(A))' \setminus \{(0, 0)\}$ we examine $\min_{y \in A} l^\top y$. We consider three cases.

As for $y \in A$, $y_2 \geq 1$ implies $y_1 \geq 0$ and $y_2 < 1$ implies $y_1 \geq 1$, we obtain for
$l_1 > l_2 \geq 0$

$$l^\top (0,1) = l_2 \le \begin{cases} l_1 \le l_1 y_1 + l_2 y_2 & \text{if } y_2 < 1 \\ l_1 y_1 + l_2 y_2 & \text{if } y_2 \ge 1 \end{cases} = l^\top y \text{ for all } y \in A.$$

For $l_2 > l_1 \ge 0$ we obtain

$$l^\top (1,0) = l_1 \le l^\top y \text{ for all } y \in A$$

and finally for $l_1 = l_2 > 0$,

$$(l_1, l_2)(1,0) = (l_1, l_2)(0,1) = l_1 \le l^\top y \text{ for all } y \in A.$$

Hence, $(0,1)$ and $(1,0)$ are weakly nondominated elements of A w.r.t. $\mathcal{D}$ but, for instance, all points of the set $\{(t,1) \mid t \in (0,1]\}$ are also weakly nondominated but cannot be found by applying the sufficient condition.

By Lemma 2.23(ii), the stated necessary condition for weakly nondominated elements w.r.t. $\mathcal{D}$ is also a necessary condition for nondominated elements w.r.t. $\mathcal{D}$. For sufficient conditions see the following theorem.

Theorem 4.6. (a) *If for some $l \in (\mathcal{D}(A))'$ the element $\bar{y} \in A$ is a unique minimal solution of $\min_{y \in A} l(y)$, i.e.*

$$l(\bar{y}) < l(y) \text{ for all } y \in A \setminus \{\bar{y}\},$$

then $\bar{y}$ is a nondominated element of A w.r.t. $\mathcal{D}$.
(b) *If for some $l \in (\mathcal{D}(A))_Y^\#$, the element $\bar{y} \in A$ is a minimal solution of $\min_{y \in A} l(y)$, i.e.*

$$l(\bar{y}) \le l(y) \text{ for all } y \in A,$$

then $\bar{y}$ is a nondominated element of A w.r.t. $\mathcal{D}$.

Proof. The proof is similar to the proof of Theorem 4.2(b).   □

See Theorem 1.20 for conditions ensuring that $(\mathcal{D}(A))^\#$ is nonempty. The main drawbacks in using linear scalarizations are that we need convexity of the set A for the necessary conditions and that $(\mathcal{D}(A))' = \bigcap_{y \in A} \mathcal{D}(y)'$ must not equal the trivial cone $\{0_{Y'}\}$.

Example 4.7. Let $Y = \mathbb{R}^2$ and

$$A := \{(y_1, y_2) \in \mathbb{R}^2 \mid \|(y_1, y_2) - (1,1)\|_2 \le 1\},$$

see Fig. 4.2, and let the cone-valued map $\mathcal{D}$ be defined as in Example 1.11 on page 8. Then $\mathcal{D}(y) \subset \mathbb{R}_+^2$ for all $y \in \mathbb{R}^2$ and $\mathcal{D}(A) = \mathbb{R}_+^2$ and thus $(\mathcal{D}(A))' = \mathbb{R}_+^2$ and

$$(\mathcal{D}(A))^\# = \{(y_1, y_2) \in \mathbb{R}^2 \mid y_1 > 0, \ y_2 > 0\} \ne \emptyset.$$

Fig. 4.2 Illustration of
Example 4.7, cf. [46]

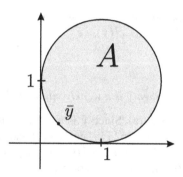

For $(1, 1)^\top \in (\mathcal{D}(A))^\#$ the point $\bar{y} = (1 - (1/\sqrt{2}), 1 - (1/\sqrt{2}))^\top$ is a minimal solution of $\min_{y \in A} l^\top y$ and hence, according to Theorem 4.6(b), $\bar{y}$ is a nondominated element of A w.r.t. $\mathcal{D}$.

As a direct consequence of the previous theorem we obtain the following existence result for nondominated elements w.r.t. a variable ordering structure.

Theorem 4.8. *Let Y be a real topological linear space and let A be a nonempty compact subset of Y. Let $(\mathcal{D}(A))^\#_{Y*}$ be nonempty. Then a nondominated element of the set A w.r.t. $\mathcal{D}$ exists.*

Proof. The result follows from the Weierstraß theorem together with Theorem 4.6(b).   □

Note that existence results for nondominated elements w.r.t. a variable ordering structure are difficult to obtain taking into account that already one element $\tilde{y} \in A$ with $y \in \{\tilde{y}\} + (\mathcal{D}(\tilde{y}) \setminus \{0_Y\})$ satisfies for y being not a nondominated element of A w.r.t. $\mathcal{D}$ and examples can be easily constructed where no nondominated element exists, see for instance Example 2.9. So, strict assumptions are necessary.

4.2   Characterization of Minimal Elements

Based on Lemma 2.15, many of the well-known scalarization results in partially ordered spaces can be applied to characterize minimal elements w.r.t. $\mathcal{D}$ with no or only few modifications. In the following, we give some of these characterizations.

Theorem 4.9. *Let $\bar{y} \in A$ and $cor(\mathcal{D}(\bar{y}))$ be nonempty.*

(a) *Let the set A be convex and let $\bar{y}$ be a weakly minimal element of A w.r.t. $\mathcal{D}$. Then a linear functional $l \in \mathcal{D}(\bar{y})' \setminus \{0_{Y'}\}$ exists with $\bar{y}$ a minimal solution of $\min_{y \in A} l(y)$, i.e.*

$$l(\bar{y}) \le l(y) \, for \, all \, y \in A.$$

(b) *If for some* $l \in \mathcal{D}(\bar{y})' \setminus \{0_{Y'}\}$ *the element* $\bar{y}$ *is a minimal solution of* $\min_{y \in A} l(y)$, *i.e.*

$$l(\bar{y}) \leq l(y) \text{ for all } y \in A,$$

then $\bar{y}$ *is a weakly minimal element of* A *w.r.t.* $\mathcal{D}$.

Proof. (a) Since $\bar{y}$ is a weakly minimal element of A w.r.t. $\mathcal{D}$ the intersection of the sets $\{\bar{y}\} - \text{cor}(\mathcal{D}(\bar{y}))$ and A is empty. Applying Theorem 4.1 and using that $\mathcal{D}(\bar{y})$ is a cone (see the proof of Theorem 4.2(a)), the assertion follows.

(b) If $\bar{y} - y \in \text{cor}(\mathcal{D}(\bar{y}))$ for any $y \in A$ then $l \in \mathcal{D}(\bar{y})' \setminus \{0_{Y'}\}$ implies by Lemma 1.7(i) $l(\bar{y}) > l(y)$. Thus, as $l(\bar{y}) \leq l(y)$ for all $y \in A$, $A \cap (\{\bar{y}\} - \text{cor}(\mathcal{D}(\bar{y}))) = \emptyset$.

$\square$

As $\mathcal{D}(y)' \supset (\mathcal{D}(A))'$ for all $y \in A$, we obtain:

Corollary 4.10. *If for some* $l \in (\mathcal{D}(A))' \setminus \{0_{Y'}\}$ *the element* $\bar{y} \in A$ *is a minimal solution of* $\min_{y \in A} l(y)$, *i.e.*

$$l(\bar{y}) \leq l(y) \text{ for all } y \in A,$$

then $\bar{y}$ *is a weakly minimal element of* A *w.r.t.* $\mathcal{D}$.

In the case of convexity of the set A, the necessary and sufficient conditions completely characterize weakly minimal elements w.r.t. a variable ordering structure. We will see that this does not hold for minimal elements w.r.t. a variable ordering structure. As any minimal element of a set w.r.t. $\mathcal{D}$ is also a weakly minimal element w.r.t. $\mathcal{D}$ (in case of a nonempty interior of the cones $\mathcal{D}(y)$ for all $y \in A$), the necessary condition given in Theorem 4.9 is also necessary for minimal elements w.r.t. a variable ordering structure. In the following theorem we give two sufficient conditions for minimal elements w.r.t. a variable ordering structure.

Theorem 4.11. (a) *If for* $\bar{y} \in A$ *and some* $l \in \mathcal{D}(\bar{y})'$ *the element* $\bar{y}$ *is a unique minimal solution of* $\min_{y \in A} l(y)$, *i.e.*

$$l(\bar{y}) < l(y) \text{ for all } y \in A \setminus \{\bar{y}\},$$

then $\bar{y}$ *is a minimal element of* A *w.r.t.* $\mathcal{D}$.

(b) *If for* $\bar{y} \in A$ *and some* $l \in \mathcal{D}(\bar{y})^{\#}$ *the element* $\bar{y}$ *is a minimal solution of* $\min_{y \in A} l(y)$, *i.e.*

$$l(\bar{y}) \leq l(y) \text{ for all } y \in A,$$

then $\bar{y}$ *is a minimal element of* A *w.r.t.* $\mathcal{D}$.

Proof. (a) If $\bar{y}$ is not a minimal element of A w.r.t. $\mathcal{D}$, then $\bar{y} - y \in \mathcal{D}(\bar{y}) \setminus \{0_Y\}$ for some $y \in A$ and as $l \in \mathcal{D}(\bar{y})'$ this implies $l(\bar{y}) \geq l(y)$ in contradiction to the assumption.

(b) If $\bar{y} - y \in \mathcal{D}(\bar{y}) \setminus \{0_Y\}$ for some $y \in A$, we get by $l \in \mathcal{D}(\bar{y})^\#$ that $l(\bar{y}) > l(y)$ in contradiction to the assumption.

$\square$

In case the variable ordering structure is given by a cone-valued map with images Bishop-Phelps cones, i.e. $\mathcal{D}(y) = C(\ell(y))$ for all $y \in Y$ for some map $\ell : Y \to Y^*$, compare Sect. 1.2, then according to Lemma 1.16(v) $\ell(y) \in C(\ell(y))^\#_{Y*} = \mathcal{D}(y)^\#_{Y*}$ for all $y \in Y$. Thus, in Theorem 4.11(b), we can choose $l := \ell(\bar{y})$.

As $\mathcal{D}(\bar{y})' \supset (\mathcal{D}(A))'$ and $\mathcal{D}(\bar{y})^\# \supset (\mathcal{D}(A))^\#$ we obtain:

Corollary 4.12. *(a) If for some $l \in (\mathcal{D}(A))'$ the element $\bar{y} \in A$ is a unique minimal solution of $\min_{y \in A} l(y)$, i.e.*

$$l(\bar{y}) < l(y) \text{ for all } y \in A \setminus \{\bar{y}\},$$

then $\bar{y}$ is a minimal element of A w.r.t. $\mathcal{D}$.

(b) If for some $l \in (\mathcal{D}(A))^\#$ the element $\bar{y} \in A$ is a minimal solution of $\min_{y \in A} l(y)$, i.e.

$$l(\bar{y}) \leq l(y) \text{ for all } y \in A,$$

then $\bar{y}$ is a minimal element of A w.r.t. $\mathcal{D}$.

We apply the above scalarization results to the set A of Example 2.10.

Example 4.13. Let Y, A and $\mathcal{D}$ be defined as in Example 2.10. Then A is convex. We examine whether $\bar{y} = (-1, 0)$ satisfies the necessary conditions of Theorem 4.9(a) for being a weakly minimal element of A w.r.t. $\mathcal{D}$. As $\mathcal{D}(\bar{y}) = \mathbb{R}^2_+$, also $\mathcal{D}(\bar{y})' = \mathbb{R}^2_+$. For $l = (1, 0)$ we obtain

$$l^T(-1, 0)^T = -1 < l^T y = y_1 \text{ for all } (y_1, y_2) \in A \setminus \{\bar{y}\}. \tag{4.1}$$

Thus, $\bar{y}$ satisfies the necessary condition of Theorem 4.9(a); and also the sufficient condition of Theorem 4.11(a).

Next we consider the element $\bar{y} = (0, -1)$. For any

$$(l_1, l_2) \in \mathcal{D}(\bar{y})' \setminus \{(0, 0)\} = \{(y_1, y_2) \in \mathbb{R}^2_+ \mid y_2 \leq y_1\} \setminus \{(0, 0)\}$$

it holds that $l_2 \leq (l_1 + l_2)/2$. We obtain for $y = (-1/\sqrt{2}, -1/\sqrt{2}) \in A$ and any $l \in \mathcal{D}(\bar{y})' \setminus \{(0, 0)\}$ that

$$l^T \bar{y} = -l_2 \geq -(l_1 + l_2)/2 > -(l_1 + l_2)/\sqrt{2} = l^T y.$$

Fig. 4.3 The set A of
Example 4.14 and, as *dotted
line*, the set
$\{y \in \mathbb{R}^2 \mid l^\top y = 1\}$ for
$l = (1, 1)$

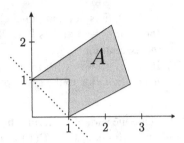

Thus $(0, -1)$ does not satisfy the necessary conditions of Theorem 4.9(a) and is thus
not a weakly minimal element of A w.r.t. $\mathcal{D}$

In the following we reconsider Example 4.5 with a nonconvex set A.

Example 4.14. Let Y, A and $\mathcal{D}$ be specified as in Example 4.5. We apply
Theorem 4.9(b) to determine weakly minimal elements of A w.r.t. $\mathcal{D}$. As
$(\mathcal{D}(A))' = \mathbb{R}^2_+ \subset (\mathcal{D}(y))'$ for all $y \in A$, according to Example 4.5 there exists
some $l \in \mathbb{R}^2_+ \subset (\mathcal{D}(0, 1))'$ with $l^\top (0, 1) \leq l^\top y$ for all $y \in A$ and also there is some
$l \in \mathbb{R}^2_+ \subset (\mathcal{D}(1, 0))'$ with $l^\top (1, 0) \leq l^\top y$ for all $y \in A$. Hence, $(0, 1)$ and $(1, 0)$
are weakly minimal elements of A w.r.t. $\mathcal{D}$.

All elements of the set

$$\{(1, t) \in \mathbb{R}^2 \mid t \in (0, 1]\} \cup \{(t, 1) \in \mathbb{R}^2 \mid t \in (0, 1)\}$$

are also weakly minimal elements of A w.r.t. $\mathcal{D}$ but cannot be found as minimal
solutions of $\min_{y \in A} l^\top y$ for some linear functional $l \in \mathbb{R}^2 \setminus \{(0, 0)\}$, see Fig. 4.3.

By comparing Corollaries 4.10 and 4.12 with Theorems 4.2(b) and 4.6, it turns
out that the same sufficient conditions simultaneously guarantee (weakly) minimal
and nondominated elements w.r.t. $\mathcal{D}$. These sufficient conditions are very strong as
they demand $(\mathcal{D}(A))' \neq \{0_{Y'}\}$.

The following characterization of minimal elements w.r.t. a variable ordering
structure is also based on a similar result for efficient elements in a partially ordered
space:

Lemma 4.15. *Let Y be a real locally convex topological linear space, $\bar{y} \in A$ and
let $\mathcal{D}(\bar{y})$ be additionally closed. Then $\bar{y}$ is a minimal element of the set A w.r.t.
$\mathcal{D}$ if and only if for every $y \in A \setminus \{\bar{y}\}$ there is a continuous linear functional
$l \in \mathcal{D}(\bar{y})^* \setminus \{0_{Y^*}\}$ with*

$$l(\bar{y}) < l(y).$$

Proof. Using Lemma 2.15 this result is a direct consequence of known results, see
for instance [94, Theorem 5.5]. $\qquad\square$

For $(\mathcal{D}(A))^{\#}$ nonempty we obtain the following existence result for minimal elements.

Theorem 4.16. *Let Y be a real topological linear space and let A be a nonempty compact subset of Y. Let $(\mathcal{D}(A))^{\#}$ be nonempty. Then a minimal element of the set A w.r.t. $\mathcal{D}$ exists.*

Proof. The result is a consequence of the Weierstraß theorem together with Corollary 4.12(b). $\qquad\square$

Example 4.17. Let Y, A and $\mathcal{D}$ be defined as in Example 4.7. According to Theorem 4.16 a minimal element of A w.r.t. $\mathcal{D}$ exists.

Indeed, for $(1, 1)^\top \in (\mathcal{D}(A))^{\#}$ the point

$$\bar{y} = (1 - (1/\sqrt{2}), 1 - (1/\sqrt{2}))^\top$$

is a minimal solution of $\min_{y \in A} l^\top y$ and hence, according to Corollary 4.12(b), a minimal element of A w.r.t. $\mathcal{D}$.

We conclude with a result on strongly minimal elements w.r.t. a variable ordering structure.

Theorem 4.18. *Let Y be a real locally convex topological linear space and let $\mathcal{D}(y)$ be additionally closed for all $y \in Y$. An element $\bar{y} \in A$ is a strongly minimal element of the set A if and only if for every $l \in \mathcal{D}(\bar{y})^*$*

$$l(\bar{y}) \leq l(y) \text{ for all } y \in A,$$

i.e. $\bar{y}$ is a minimal solution of the scalar-valued problem $\min_{y \in A} l(y)$.

Proof. Since $\mathcal{D}(\bar{y})$ is closed and convex it holds $\mathcal{D}(\bar{y}) = \{y \in Y \mid l(y) \geq 0 \text{ for all } l \in \mathcal{D}(\bar{y})^*\}$ by Lemma 1.7(ii). The element $\bar{y} \in A$ is a strongly minimal element of A if and only if $A - \{\bar{y}\} \subset \mathcal{D}(\bar{y})$, i.e. if and only if it holds for all $y \in A$ that $l(y - \bar{y}) \geq 0$ for all $l \in D(\bar{y})^*$. $\qquad\square$

Example 4.19. Let Y, A and $\mathcal{D}$ be defined as in Example 2.14. Then for $\bar{y} = (0, 0)$ it is $\mathcal{D}(\bar{y})^* = \mathbb{R}_+^2$ and for all $l \in \mathcal{D}(\bar{y})^*$ it holds $l^\top \bar{y} = 0$. Because of $A \subset \mathbb{R}_+^2$, $l^\top y \geq 0$ for all $y \in A$. Hence, $\bar{y} = (0, 0)$ is a strongly minimal element of A w.r.t. $\mathcal{D}$ according to Theorem 4.18.

4.3 Notes on the Literature

The linear scalarization results presented here are based on the fact that a functional is chosen which is monotonically increasing w.r.t. the binary relation. A functional $l: Y \to \mathbb{R}$ is denoted *monotonically increasing* if for all $y^1, y^2 \in Y$

$$y^1 \leq y^2 \quad \text{implies} \quad l(y^1) \leq l(y^2).$$

Here, we consider $\leq = \leq_1$ or $\leq = \leq_2$. It is easy to show that if $\bar{y} \in A$ is a unique minimal solution of

$$\min_{y \in A} l(y)$$

with l a monotonically increasing functional, then it is an optimal (w.r.t. $\leq$) element of A w.r.t. $\mathcal{D}$. If $l \in (\mathcal{D}(A))'$, then l is monotonically increasing w.r.t. both relations $\leq_1$ and $\leq_2$, which leads to the results of Theorem 4.6(a) and Corollary 4.12(a).

Scalarization results derived from the monotonicity of the functionals were already given by Luc in [112] and he considered in this context also the nonlinear scalarization functional denoted as translative or Gerstewitz functional which we will examine in Sect. 5.2. Monotonicity arguments are also used for instance by Jahn in his book [94].

In the finite dimensional linear space $Y = \mathbb{R}^m$ partially ordered by $K = \mathbb{R}^m_+$, linear scalarizations for $l \in K' = \mathbb{R}^m_+$ are also known as weighted sum approach with the components l_i, $i = 1, \ldots, m$, of the vector l denoted as weights, see for instance the books by Ehrgott [39], Eichfelder [42] or Miettinen [116]. Linear scalarizations are used in numerical approaches for determining approximations of the set of efficient elements for instance in the book by Eichfelder [42], or for formulating existence and duality results, see Chaps. 6 and 8 in the book by Jahn [94].

For vector optimization problems w.r.t. a variable ordering structure linear scalarizations have first been proposed by Engau [57, 58] for the notion of minimal elements w.r.t. an ordering map $\mathcal{D}$ defined on $\mathbb{R}^m$ with $\mathbb{R}^m_+ \subset \mathcal{D}(y)$ for all y. For both optimality concepts characterizations based on linear scalarizations have been given in [46, 47]. They were also extensively studied by Gebhardt in her diploma thesis [61]. There, many examples are presented illustrating the relevance of the necessary and sufficient conditions for (weakly) nondominated and (weakly) minimal elements w.r.t. a variable ordering structure.

Linear functionals are used in partially ordered linear spaces also for characterizing properly efficient elements. These results can be generalized to variable ordering structures for characterizing properly minimal and nondominated elements. For results we refer to [55].

Note that one may also define a scalarization functional by choosing for each $y \in Y$ an element $\ell(y) \in \mathcal{D}(y)'$, i.e. by defining a map $\ell: Y \to Y'$ with $\ell(y) \in \mathcal{D}(y)' \setminus \{0_{Y'}\}$ for all $y \in Y$. Then to any $\bar{y} \in A$ a functional $s_{\bar{y}}: Y \to \mathbb{R}$ is defined by

$$s_{\bar{y}}(y) = \ell(y)(y - \bar{y}) = \ell(y)(y) - \ell(y)(\bar{y}) .$$

Based on this functional also a complete characterization of (weakly) nondominated elements is possible, see [55]. However, even in case $\ell: Y \to Y'$ is a linear map, $s_{\bar{y}}$ is in general not a linear functional.

Chapter 5
Nonlinear Scalarizations

Linear scalarizations, as discussed in the previous chapter, are in many cases, for instance if the considered set is nonconvex or if the cone $\mathcal{D}(A)$ is not pointed, not an adequate tool for characterizing optimal elements. Moreover, they do not allow a complete characterization of (weakly) nondominated elements. Therefore, we next examine nonlinear scalarizations. We study parameter dependent nonlinear functionals based on which a complete characterization of optimal elements w.r.t. a variable ordering structure is possible. We examine these functionals on convexity in view of using the functionals to formulate sufficient optimality conditions for optimal solutions of vector optimization problems in form of Fermat and Lagrange multiplier rule. Especially for the case of an ordering map $\mathcal{D}$ with images Bishop-Phelps cones strong results can be derived by considering a nonlinear scalarization functional using the special structure of Bishop-Phelps cones. For these more special scalarization functionals we refer to Chap. 6.

In this chapter, we assume Y to be a real topological linear space equipped with a variable ordering structure defined by the ordering map $\mathcal{D}: Y \rightarrow 2^Y$ with $\mathcal{D}(y)$ a nontrivial closed pointed convex cone for all $y \in Y$, and A to be a nonempty subset of Y.

5.1 Signed Distance Functional

In this section we assume additionally that $(Y, \| \cdot \|)$ is a real normed space. Based on the distance function $d_S: Y \rightarrow \mathbb{R} \cup \{+\infty\}$ for some set $S \subset Y$, defined by

$$d_S(y) := \inf\{\|y - s\| \mid s \in S\} \quad \text{for all } y \in Y,$$

with $d_\emptyset(y) := \infty$ for all $y \in Y$, Hiriart-Urruty introduced in [84, 85] the function $\Delta_S: Y \rightarrow \mathbb{R} \cup \{\pm\infty\}$,

G. Eichfelder, *Variable Ordering Structures in Vector Optimization*, Vector Optimization,
DOI 10.1007/978-3-642-54283-1_5, © Springer-Verlag Berlin Heidelberg 2014

$$\Delta_S(y) := d_S(y) - d_{Y\setminus S}(y) = \begin{cases} d_S(y) & \text{for } y \in Y \setminus S, \\ -d_{Y\setminus S}(y) & \text{for } y \in S. \end{cases} \qquad (5.1)$$

This function, denoted as *signed distance function* or as *Hiriart-Urruty function*, has several useful properties:

Lemma 5.1 ([84, 110, 160]). *Let S be a nonempty subset of Y with $S \neq Y$.*

(i) $\Delta_S(y) \in \mathbb{R}$ *for all $y \in Y$ and Δ_S is Lipschitz continuous with constant 1.*

(ii) *If S is convex, then Δ_S is convex, and if Δ_S is convex, then $cl(S)$ is convex.*

(iii) $\Delta_{Y\setminus S} = -\Delta_S$.

(iv) $\Delta_S(y) < 0$ *if and only if $y \in int(S)$;*
$\Delta_S(y) = 0$ *if and only if $y \in \partial S$;*
$\Delta_S(y) > 0$ *if and only if $y \in int(Y \setminus S) = Y \setminus cl(S)$.*

(v) *If S is a cone, then Δ_S is positively homogeneous.*

(vi) *If S is convex with nonempty interior, then*

$$\Delta_S(y) = \sup_{y^* \in B^*} \inf_{s \in S} y^*(y - s)$$

with $B^* = \{y^* \in Y^* \mid \|y^*\|_* = 1\}$.

(vii) *Let S_1, S_2 be nonempty closed subsets of Y with $S_1, S_2 \neq Y$. Then*

$$S_1 \subset S_2 \Leftrightarrow \Delta_{S_2}(y) \leq \Delta_{S_1}(y) \text{ for all } y \in Y.$$

Note that if S is closed, then $\Delta_S(y) > 0$ if and only if $y \notin S$. For scalarization results for efficient elements in a partially ordered space the set S is replaced in (5.1) by $-K$ with K the closed pointed convex cone defining the partial ordering in the space Y. Under these assumptions the functional is Lipschitz continuous, convex and sublinear [160]. Further, for all $y^1, y^2 \in Y$,

$$y^2 \in \{y^1\} + K \text{ implies } \Delta_{-K}(y^1) \leq \Delta_{-K}(y^2)$$

and

$$y^2 \in \{y^1\} + int(K) \text{ implies } \Delta_{-K}(y^1) < \Delta_{-K}(y^2),$$

i.e. Δ_{-K} is strictly monotonically increasing on Y w.r.t. the partial ordering introduced by K [110, 160].

We illustrate in $Y = \mathbb{R}^m$ the functional defined in (5.1) with $-S = K = \mathbb{R}^m_+$, cf. [160]:

Example 5.2. Let $Y = \mathbb{R}^m$ and $K = \mathbb{R}^m_+$, i.e. $S = -\mathbb{R}^m_+$.

(a) Assuming the Euclidean norm we obtain $d_{-K}(y) = \|y^+\|_2$ with y^+ defined by $y_i^+ := \max\{0, y_i\}$ for all $i = 1, \ldots, m$, and

$$d_{Y \setminus -K}(y) = \begin{cases} 0 & \text{if } y_i > 0 \text{ for some } i \in \{1, \ldots, m\}, \\ -\max_{i \in \{1, \ldots, m\}} y_i & \text{if } y_i \leq 0 \text{ for all } i = 1 \ldots, m. \end{cases}$$

Thus

$$\Delta_{-K}(y) = \begin{cases} \|y^+\|_2 & \text{if } y \notin -K, \\ \max_{i \in \{1, \ldots, m\}} y_i & \text{if } y \in -K. \end{cases}$$

(b) Assuming the Maximum norm $\|y\|_\infty := \max_{i \in \{1, \ldots, m\}} |y_i|$ for all $y \in Y$, we obtain

$$\Delta_{-K}(y) = \max_{i \in \{1, \ldots, m\}} y_i.$$

The representation of the functional is also simpler for other linear spaces and ordering cones K:

Example 5.3. For $Y = C(S)$ the real normed space of continuous functionals defined over some compact set S with the supremum norm, we obtain for the natural ordering cone

$$K = \{f \in C(S) \mid f(x) \geq 0 \text{ for all } x \in S\}$$

that the signed-distance functional is given by [160]

$$\Delta_{-K}(f) = \max_{x \in S} f(x) \quad \text{for all } f \in C(S).$$

Efficient elements in a partially ordered space Y, ordered by K as given above, can be characterized as follows [160]: an element $\bar{y} \in A$ is an efficient element if and only if $\bar{y}$ is a unique minimal solution of

$$\min_{y \in A} \Delta_{-K}(y - \bar{y}),$$

i.e. if and only if

$$\Delta_{-K}(y - \bar{y}) > \Delta_{-K}(0_Y) = 0 \text{ for all } y \in A \setminus \{\bar{y}\}.$$

Using Lemma 2.15 we directly obtain characterizations of minimal elements w.r.t. a variable ordering structure. We refer to Sect. 5.1.2. First we start by modifying the functional in (5.1) for being able to characterize also nondominated elements w.r.t. a variable ordering structure.

5.1.1 Characterization of Nondominated Elements

By allowing the set S in (5.1) to vary dependently on the actual element y, also a complete characterization of nondominated elements w.r.t. the ordering map $\mathcal{D}$ is possible. Thus we consider additionally the function $\sigma_{\bar{y}} \colon Y \to \mathbb{R} \cup \{\pm\infty\}$,

$$\sigma_{\bar{y}}(y) := \Delta_{-\mathcal{D}(y)}(y - \bar{y})$$
$$= d_{-\mathcal{D}(y)}(y - \bar{y}) - d_{Y \setminus (-\mathcal{D}(y))}(y - \bar{y}) \text{ for all } y \in Y$$

for some given element $\bar{y} \in Y$.

We collect some basic properties of $\sigma_{\bar{y}}$. Recall that we have assumed $\mathcal{D}(y)$ to be closed for all $y \in Y$

Lemma 5.4. *Let $\bar{y} \in Y$ be given.*

(i) $\sigma_{\bar{y}}(y) \in \mathbb{R}$ *for all $y \in Y$.*
(ii) $\sigma_{\bar{y}}(y) < 0$ *if and only if $\bar{y} - y \in int(\mathcal{D}(y))$;*
   $\sigma_{\bar{y}}(y) \leq 0$ *if and only if $\bar{y} - y \in \mathcal{D}(y)$;*
   $\sigma_{\bar{y}}(\bar{y}) = 0.$

Proof. (i) Follows by Lemma 5.1(i) as $\sigma_{\bar{y}}(y) = \Delta_{-\mathcal{D}(y)}(y - \bar{y})$.
(ii) Follows by Lemma 5.1(iv) and because $0_Y \in \partial(-\mathcal{D}(\bar{y}))$ and thus $\sigma_{\bar{y}}(\bar{y}) = \Delta_{-\mathcal{D}(\bar{y})}(0_Y) = 0$.

$\square$

For an examination whether $\sigma_{\bar{y}}$ is convex on Y, i.e. whether

$$\sigma_{\bar{y}}(\lambda y^1 + (1 - \lambda) y^2) \leq \lambda \sigma_{\bar{y}}(y^1) + (1 - \lambda) \sigma_{\bar{y}}(y^2) \tag{5.2}$$

holds for all $y^1, y^2 \in Y$, $\lambda \in (0, 1)$, we first use the definition of the functional to state that (5.2) is equivalent to

$$\Delta_{-\mathcal{D}(y^\lambda)}(y^\lambda - \bar{y}) \leq \lambda \Delta_{-\mathcal{D}(y^1)}(y^1 - \bar{y}) + (1 - \lambda) \Delta_{-\mathcal{D}(y^2)}(y^2 - \bar{y}) \tag{5.3}$$

with

$$y^\lambda := \lambda y^1 + (1 - \lambda) y^2.$$

As $-\mathcal{D}(y^\lambda)$ is a closed convex cone, we obtain due to the sublinearity of $\Delta_{-\mathcal{D}(y^\lambda)}$ [160]

$$\Delta_{-\mathcal{D}(y^\lambda)}(y^\lambda - \bar{y}) \leq \Delta_{-\mathcal{D}(y^\lambda)}(\lambda(y^1 - \bar{y})) + \Delta_{-\mathcal{D}(y^\lambda)}((1 - \lambda)(y^2 - \bar{y}))$$
$$= \lambda \Delta_{-\mathcal{D}(y^\lambda)}(y^1 - \bar{y}) + (1 - \lambda) \Delta_{-\mathcal{D}(y^\lambda)}(y^2 - \bar{y}). \tag{5.4}$$

Comparing (5.4) with (5.3) yields that convexity of $\sigma_{\bar{y}}$ is given if

$$\Delta_{-\mathcal{D}(y^\lambda)}(y^i - \bar{y}) \leq \Delta_{-\mathcal{D}(y^i)}(y^i - \bar{y}), \qquad i = 1, 2.$$

By Lemma 5.1(vii) this holds if

$$\mathcal{D}(y^i) \subset \mathcal{D}(y^\lambda) = \mathcal{D}(\lambda\, y^1 + (1 - \lambda)\, y^2), \qquad i = 1, 2.$$

This, however, implies according to Lemma 3.11 that $\mathcal{D}$ is constant. Thus, the functional $\sigma_{\bar{y}}$ can in general not be assumed to be convex.

The scalarization results for (weakly) nondominated elements are summed up in the next theorem:

Theorem 5.5. *(a)* $\bar{y} \in A$ *is a nondominated element of A w.r.t. the ordering map* $\mathcal{D}$ *if and only if* $\bar{y}$ *is a unique minimal solution of* $\min_{y \in A} \sigma_{\bar{y}}(y)$, *i.e. if and only if*

$$\sigma_{\bar{y}}(y) > \sigma_{\bar{y}}(\bar{y}) = 0 \text{ for all } y \in A \setminus \{\bar{y}\}.$$

(b) Let $int(\mathcal{D}(y)) \neq \emptyset$ *for all* $y \in A$. $\bar{y} \in A$ *is a weakly nondominated element of A w.r.t. the ordering map* $\mathcal{D}$ *if and only if* $\bar{y}$ *is a minimal solution of* $\min_{y \in A} \sigma_{\bar{y}}(y)$, *i.e. if and only if*

$$\sigma_{\bar{y}}(y) \geq \sigma_{\bar{y}}(\bar{y}) = 0 \text{ for all } y \in A.$$

Proof. According to Lemma 5.4(ii) $\sigma_{\bar{y}}(\bar{y}) = 0$.

(a) The element $\bar{y}$ is a nondominated element of A w.r.t. $\mathcal{D}$ if and only if $\bar{y} - y \notin \mathcal{D}(y)$ for all $y \in A \setminus \{\bar{y}\}$ and by Lemma 5.4(ii) this is equivalent to $\sigma_{\bar{y}}(y) > 0$ for all $y \in A \setminus \{\bar{y}\}$.

(b) The element $\bar{y}$ is a weakly nondominated element of A w.r.t. $\mathcal{D}$ if and only if $\bar{y} - y \notin int\mathcal{D}(y)$ for all $y \in A$ and by Lemma 5.4(ii) this is equivalent to $\sigma_{\bar{y}}(y) \geq 0$ for all $y \in A$.

$\square$

The following example illustrates the result of Theorem 5.5. It also shows that calculating $\sigma_{\bar{y}}(y)$ is closely related to checking whether $\bar{y} - y \in \mathcal{D}(y)$. Using numerical methods for calculating $\sigma_{\bar{y}}(y)$, scalar-valued optimization problems have to be solved for evaluating the distance function.

Example 5.6 ([61]). Consider the Euclidean space $Y = \mathbb{R}^2$ and

$$A := \{(y_1, y_2) \in \mathbb{R}^2 \mid y_1 + y_2 \geq 2, -\frac{1}{2}y_1 + y_2 \leq 2, \frac{2}{3} \leq y_1 \leq 3, y_2 \geq 1\},$$

see Fig. 5.1.

Fig. 5.1 Set A of
Example 5.6, cf. [61]

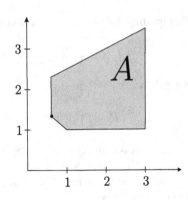

We define by A^s a subset of the boundary of A:

$$A^s := \{(y_1, y_2) \in \mathbb{R}^2 \mid (y_1 + y_2 = 2 \land \tfrac{2}{3} < y_1 < 1)$$
$$\lor (1 \leq y_1 \leq 3 \land y_2 = 1)\}.$$

The variable ordering structure is given by $\mathcal{D} : \mathbb{R}^2 \to 2^{\mathbb{R}^2}$ with

$$\mathcal{D}(y) := \begin{cases} \mathbb{R}^2_+ & \text{for } y \in \mathbb{R}^2 \setminus A^s, \\ \{(y_1, y_2) \in \mathbb{R}^2 \mid y_2 \geq 0, \ y_1 + y_2 \geq 0\} & \text{for } y \in A^s. \end{cases}$$

We examine the point $\bar{y} = (2/3, 4/3) \in A \setminus A^s$. Let $y \in A \setminus \{\bar{y}\}$ be given. We use Lemma 5.4(ii) and check whether $\bar{y} - y \in \mathcal{D}(y)$ or $\bar{y} - y \in \text{int}(\mathcal{D}(y))$.

If $y_2 > 4/3$, then $\bar{y}_2 - y_2 < 0$ and $\bar{y} - y \notin \mathcal{D}(y) = \mathbb{R}^2_+$, i.e. $\sigma_{\bar{y}}(y) > 0$.

If $y_2 \leq 4/3$, then, because of $y \neq \bar{y}$, $y_1 > 2/3$. If $y \in \mathbb{R}^2 \setminus A^s$ then we have by $\bar{y}_1 - y_1 < 0$ already $\bar{y} - y \notin \mathcal{D}(y) = \mathbb{R}^2_+$, i.e. $\sigma_{\bar{y}}(y) > 0$. Else, we further have $\bar{y}_2 - y_2 \geq 0$ and as $y_1 + y_2 \geq 2$

$$(\bar{y}_1 - y_1) + (\bar{y}_2 - y_2) = 2 - (y_1 + y_2) \leq 0.$$

Thus $\bar{y} - y \notin \text{int}(\mathcal{D}(y))$ and $\sigma_{\bar{y}}(y) \geq 0$.

Summarizing this, we have

$$\sigma_{\bar{y}}(y) \geq 0 \ \text{ for all } y \in A$$

and $\bar{y}$ is according to Theorem 5.5(b) a weakly nondominated element of A w.r.t. $\mathcal{D}$.

For $y = (1, 1) \in A^s$ we have $\sigma_{\bar{y}}(1, 1) = \Delta_{-\mathcal{D}(1,1)}(1/3, -1/3)$ and $(1/3, -1/3) \in \partial(-\mathcal{D}(1, 1))$. By Lemma 5.1(iv), $\sigma_{\bar{y}}(1, 1) = 0$, i.e. $\bar{y}$ is not a unique minimal solution of $\min_{y \in A} \sigma_{\bar{y}}(y)$ and hence not a nondominated element of A w.r.t. the ordering map $\mathcal{D}$ by Theorem 5.5(a).

For the results in Theorem 5.5 no convexity of the set A is assumed. For an example illustrating the results for a nonconvex set A we refer to [61, p. 70].

5.1.2 Characterization of Minimal Elements

Using the functional in (5.1) we can also completely characterize minimal elements of some set A w.r.t. an ordering map $\mathcal{D}$, as an element $\bar{y}$ is a minimal element of A w.r.t. $\mathcal{D}$ if and only if it is an efficient element of A in the space Y partially ordered by $K := \mathcal{D}(\bar{y})$, cf. Lemma 2.15. Hence we use the parameter-dependent functional $\tau_{\bar{y}} : Y \to \mathbb{R}$,

$$\tau_{\bar{y}}(y) := \Delta_{-\mathcal{D}(\bar{y})}(y - \bar{y}) \text{ for all } y \in Y$$

with $\bar{y} \in Y$ to completely characterize minimal elements w.r.t. the ordering map $\mathcal{D}$. Note that according to Lemma 5.1 the functional $\tau_{\bar{y}}$ is Lipschitz continuous, convex and sublinear.

Theorem 5.7. (a) $\bar{y} \in A$ is a minimal element of A w.r.t. the ordering map $\mathcal{D}$ if and only if $\bar{y}$ is a unique minimal solution of $\min_{y \in A} \tau_{\bar{y}}(y)$, i.e. if and only if

$$\tau_{\bar{y}}(y) > \tau_{\bar{y}}(\bar{y}) = 0 \text{ for all } y \in A \setminus \{\bar{y}\}.$$

(b) Let $int(\mathcal{D}(\bar{y})) \neq \emptyset$ for some element $\bar{y} \in A$. Then $\bar{y}$ is a weakly minimal element of A w.r.t. the ordering map $\mathcal{D}$ if and only if $\bar{y}$ is a minimal solution of $\min_{y \in A} \tau_{\bar{y}}(y)$, i.e. if and only if

$$\tau_{\bar{y}}(y) \geq \tau_{\bar{y}}(\bar{y}) = 0 \text{ for all } y \in A.$$

Proof. As $\mathcal{D}(\bar{y})$ is a pointed cone, $0_Y \in -\partial\mathcal{D}(\bar{y})$ and by Lemma 5.1(iv) $\tau_{\bar{y}}(\bar{y}) = \Delta_{-\mathcal{D}(\bar{y})}(0_Y) = 0$.

(a) The element $\bar{y}$ is a minimal element of A w.r.t. $\mathcal{D}$ if and only if $y - \bar{y} \notin -\mathcal{D}(\bar{y})$ for all $y \in A \setminus \{\bar{y}\}$ and by Lemma 5.1(iv) this is equivalent to $\tau_{\bar{y}}(y) = \Delta_{-\mathcal{D}(\bar{y})}(y - \bar{y}) > 0$ for all $y \in A \setminus \{\bar{y}\}$.

(b) The element $\bar{y}$ is a weakly minimal element of A w.r.t. $\mathcal{D}$ if and only if $y - \bar{y} \notin -int(\mathcal{D}(\bar{y}))$ for all $y \in A$ and by Lemma 5.1(iv) this is equivalent to $\tau_{\bar{y}}(y) = \Delta_{-\mathcal{D}(\bar{y})}(y - \bar{y}) \geq 0$ for all $y \in A$.

□

Example 5.8. We reconsider Example 5.6 and use Theorem 5.7 for determining whether $\bar{y} = (2/3, 4/3)$ is a (weakly) minimal element of A w.r.t. $\mathcal{D}$. As $\mathcal{D}(\bar{y}) = \mathbb{R}^2_+$, using Example 5.2, we need to evaluate

$$\tau_{\bar{y}}(y) = \Delta_{-\mathbb{R}^2_+}(y - \bar{y}) = \begin{cases} \|(y - \bar{y})^+\|_2 & \text{if } \bar{y} - y \notin \mathbb{R}^2_+, \\ \max_{i \in \{1,2\}}(y_i - \bar{y}_i) & \text{if } \bar{y} - y \in \mathbb{R}^2_+. \end{cases}$$

Let $y \in A \setminus \{\bar{y}\}$ be given. If $y_2 > 4/3$ then $y_2 - \bar{y}_2 > 0$ and $\bar{y} - y \notin \mathbb{R}_+^2$. Further, $y_1 \geq 2/3$ and $y_1 - \bar{y}_1 \geq 0$. Thus

$$\tau_{\bar{y}}(y) = \|(y - \bar{y})^+\|_2 = (y_1 - \bar{y}_1)^2 + (y_2 - \bar{y}_2)^2 \geq (y_2 - \bar{y}_2)^2 > 0.$$

If $y_2 \leq 4/3$ then, because of $y \neq \bar{y}$, $y_1 > 2/3$, $y_1 - \bar{y}_1 > 0$ and thus $\bar{y} - y \notin \mathbb{R}_+^2$. As $y_2 - \bar{y}_2 \leq 0$,

$$\tau_{\bar{y}}(y) = \|(y - \bar{y})^+\|_2 = (y_1 - \bar{y}_1)^2 > 0.$$

According to Theorem 5.7(a), $\bar{y}$ is a minimal and thus also a weakly minimal element of A w.r.t. $\mathcal{D}$.

5.2 Translative Functional

Allowing two parameters $a \in Y$ and $r \in Y$ the following nonlinear scalarization function $\psi_{a,r}: Y \to \mathbb{R} \cup \{\pm\infty\}$ was studied extensively for instance in vector optimization with a partially ordered space ordered by some nonempty convex cone $K \subset Y$ by Gerstewitz (Tammer) and colleagues in [64, 65] and by Pascoletti and Serafini in [128]:

$$\psi_{a,r}(y) := \inf\{t \in \mathbb{R} \mid a + t\,r - y \in K\} \text{ for all } y \in Y. \tag{5.5}$$

For a detailed survey on the usage of this functional in various contexts we refer to [80] and for an illustration see Fig. 5.2.

The following lemma collects some properties [71, Theorem 2.3.1, Corollary 2.3.5], [31, Proposition 2.1], [143] of the function.

Lemma 5.9. *Let $K \subset Y$ be a nontrivial closed pointed convex cone and $r \in K$.*

(i) *The function $\psi_{0_Y,r}$ is subadditive,*

$$\psi_{0_Y,r}(\lambda\,y) = \lambda\psi_{0_Y,r}(y) \qquad \text{for all } y \in Y, \ \lambda > 0,$$

$$\psi_{0_Y,r}(y + \lambda\,r) = \psi_{0_Y,r}(y) + \lambda \text{ for all } y \in Y, \ \lambda \in \mathbb{R},$$

and

$$-K = \{y \in Y \mid \psi_{0_Y,r}(y) \leq 0\}.$$

(ii) *The function $\psi_{a,r}$ is lower semicontinuous and convex.*

(iii) *If, additionally, $r \in \text{int}(K)$, then $\psi_{0_Y,r}$ is sublinear, the function $\psi_{a,r}$ is continuous and finite valued and*

$$\psi_{a,r}(y) = \min\{t \in \mathbb{R} \mid a + t\,r - y \in K\} \text{ for all } y \in Y.$$

Fig. 5.2 Illustration of the
function $\psi_{a,r}$ with
$\bar{t} = \psi_{a,r}(y)$

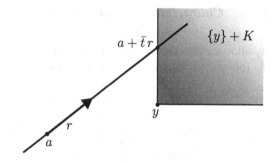

Fig. 5.3 Illustration of the
scalar-valued optimization
problem (5.6)

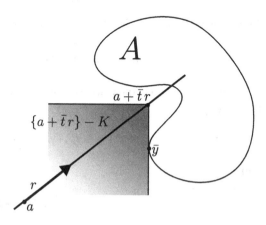

We assume in the following that the infima in the definitions of the discussed
scalarization functions are attained. For characterizing minimal elements of some
set A w.r.t. a variable ordering structure we consider for special cones K the scalar-
valued optimization problem

$$\min_{y \in A} \psi_{a,r}(y)$$

with appropriate parameters a and r, see Sect. 5.2.2. Hence we solve

$$\text{Minimize} \quad t \quad \text{subject to} \quad y \in \{a + t\,r\} - K \text{ and } t \in \mathbb{R}, \ y \in A \qquad (5.6)$$

with minimal solution $(\bar{t}, \bar{y})$—assuming a solution exists. This is for instance the
case if A is compact and $r \in \text{int}(K)$. Further conditions ensuring that are given
in [151, Chap. 6] and [153]. By solving (5.6) the cone $-K$ is moved along the ray
$\{a + t\,r \mid t \in \mathbb{R}\}$ till the cone intersects the set A. For an illustration see Fig. 5.3.

In the next two subsections, we use the scalarization functional $\psi_{a,r}$ as well as a
modification of it for characterizing minimal and nondominated elements of a vector
optimization problem w.r.t. a variable ordering structure.

5.2.1 Characterization of Nondominated Elements

By allowing also the set K to vary dependently on y we get to the ordering map $\mathcal{D}$ the following scalarization function $\chi_{a,r}: Y \to \mathbb{R} \cup \{\pm\infty\}$,

$$\chi_{a,r}(y) := \inf\{t \in \mathbb{R} \mid a + t\,r - y \in \mathcal{D}(y)\} \text{ for all } y \in Y.$$

Choosing $r \in \left(\bigcap_{y \in A} \mathcal{D}(y)\right) \setminus \{0_Y\}$ and $a := \bar{y}$, (weakly) nondominated elements w.r.t. a variable ordering structure can be characterized, see Theorem 5.11. From Lemma 5.9 we obtain the following property of this scalarization function:

Corollary 5.10. *Let* $r \in int\left(\bigcap_{y \in A} \mathcal{D}(y)\right)$. *Then*

$$\chi_{a,r}(y) = \min\{t \in \mathbb{R} \mid a + t\,r - y \in \mathcal{D}(y)\} \text{ for all } y \in Y.$$

To examine $\chi_{a,r}$ on convexity we need to check whether for all $y^1, y^2 \in Y$ and all $\lambda \in (0,1)$

$$\chi_{a,r}(y^\lambda) \leq \lambda\, \chi_{a,r}(y^1) + (1 - \lambda)\, \chi_{a,r}(y^2)$$

with

$$y^\lambda := \lambda\, y^1 + (1 - \lambda)\, y^2.$$

We assume $r \in int\left(\bigcap_{y \in A} \mathcal{D}(y)\right)$ and set $t^i := \chi_{a,r}(y^i), i = 1, 2$. Then

$$a + t^1 r - y^1 \in \mathcal{D}(y^1) \qquad \text{and} \qquad a + t^2 r - y^2 \in \mathcal{D}(y^2). \qquad (5.7)$$

We need to check that $\chi_{a,r}(y^\lambda) \leq \lambda\, t^1 + (1 - \lambda)\, t^2$, i.e. that

$$a + (\lambda\, t^1 + (1 - \lambda) t^2)\, r - y^\lambda \in \mathcal{D}(y^\lambda).$$

This is equivalent to

$$\lambda\, (a + t^1 r - y^1) + (1 - \lambda)\, (a + t^2 r - y^2) \in \mathcal{D}(y^\lambda). \qquad (5.8)$$

By (5.7) we have

$$\lambda\, (a + t^1 r - y^1) + (1 - \lambda)\, (a + t^2 r - y^2) \in \lambda\, \mathcal{D}(y^1) + (1 - \lambda)\, D(y^2). \qquad (5.9)$$

Comparing (5.9) with (5.8) yields that convexity of $\chi_{a,r}$ is given if

$$\lambda\, D(y^1) + (1 - \lambda)\, D(y^2) \subset D(\lambda\, y^1 + (1 - \lambda)\, y^2) \ \forall\, \lambda \in (0, 1),\ y^1, y^2 \in Y,$$

i.e. if $\mathcal{D}$ is convex. However, according to Corollary 3.5 this assumption is equivalent to $\mathcal{D}$ being constant. Thus, the functional $\chi_{a,r}$ can in general not be assumed to be convex.

We sum up necessary and sufficient conditions for (weakly) nondominated elements of some set A w.r.t. the ordering map $\mathcal{D}$:

Theorem 5.11. *Let $\bar{y} \in A$.*

(a) *Let $r \in \left(\bigcap_{y \in A} \mathcal{D}(y) \right) \setminus \{0_Y\}$. $\bar{y}$ is a nondominated element of A w.r.t. the ordering map $\mathcal{D}$ if and only if $\bar{y}$ is a unique minimal solution of $\min_{y \in A} \chi_{\bar{y},r}(y)$, i.e. if and only if*

$$\chi_{\bar{y},r}(y) > \chi_{\bar{y},r}(\bar{y}) = 0 \text{ for all } y \in A \setminus \{\bar{y}\}.$$

(b) *Let $r \in \text{int}\left(\bigcap_{y \in A} \mathcal{D}(y) \right)$. $\bar{y}$ is a weakly nondominated element of A w.r.t. the ordering map $\mathcal{D}$ if and only if $\bar{y}$ is a minimal solution of $\min_{y \in A} \chi_{\bar{y},r}(y)$, i.e. if and only if*

$$\chi_{\bar{y},r}(y) \geq \chi_{\bar{y},r}(\bar{y}) = 0 \text{ for all } y \in A.$$

Proof. As $\mathcal{D}(\bar{y})$ is a pointed convex cone and because of $r \in \mathcal{D}(\bar{y}) \setminus \{0_Y\}$ it holds $\chi_{\bar{y},r}(\bar{y}) = 0$.

(a) First assume $\bar{y}$ to be a nondominated element of A w.r.t. $\mathcal{D}$. If $\bar{y}$ is not a unique minimal solution of $\min_{y \in A} \chi_{\bar{y},r}(y)$ then there exists some $t \in \mathbb{R}, t \leq 0$, and some $y \in A \setminus \{\bar{y}\}$ such that

$$\bar{y} + t\,r - y \in \mathcal{D}(y).$$

Because of $t\,r \in -\mathcal{D}(y)$ and as $\mathcal{D}(y)$ is a convex cone, this implies $\bar{y} \in \{y\} + (\mathcal{D}(y) \setminus \{0_Y\})$ in contradiction to $\bar{y}$ a nondominated element of A w.r.t. $\mathcal{D}$.

Next assume $\bar{y}$ is a unique minimizer of $\chi_{\bar{y},r}$ over A. If there is some $y \in A \setminus \{\bar{y}\}$ with $\bar{y} \in \{y\} + \mathcal{D}(y)$ then this implies $\bar{y} + 0 \cdot r - y \in \mathcal{D}(y)$, i.e. $\chi_{\bar{y},r}(y) \leq 0$, which is a contradiction.

(b) First assume $\bar{y}$ to be weakly nondominated of A w.r.t. $\mathcal{D}$. If $\bar{y}$ is not a minimal solution of $\min_{y \in A} \chi_{\bar{y},r}(y)$ then there exists some $t \in \mathbb{R}, t < 0$, and some $y \in A \setminus \{\bar{y}\}$ such that $\bar{y} + t\,r - y \in \mathcal{D}(y)$. Because of $t\,r \in -\text{int}(\mathcal{D}(y))$ and as $\mathcal{D}(y)$ is a convex cone, this implies by Lemma 1.7(iv) $\bar{y} \in \{y\} + \text{int}(\mathcal{D}(y))$ in contradiction to $\bar{y}$ a weakly nondominated element of A w.r.t. $\mathcal{D}$.

Next assume $\bar{y}$ to be a minimizer of $\chi_{\bar{y},r}$ over A. If there is some $y \in A$ with $\bar{y} - y \in \text{int}(\mathcal{D}(y))$ then this implies that there exists some $t < 0$ such that $(\bar{y} - y) + t\,r \in \mathcal{D}(y)$, i.e. $\bar{y} + t\,r - y \in \mathcal{D}(y)$ and hence $\chi_{\bar{y},r}(y) < 0$, which is a contradiction.

$\square$

Fig. 5.4 Illustration of
Example 5.12, cf. [47]

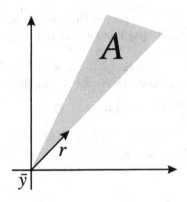

We illustrate this result by an example.

Example 5.12. Let $Y = \mathbb{R}^2$, $\mathcal{D}$ and A be given as in Example 2.14, i.e.

$$A := \{(y_1, y_2) \in \mathbb{R}^2 \mid y_1 \leq y_2 \leq 2y_1\}$$

and

$$\mathcal{D}(y_1, y_2) := \begin{cases} \mathbb{R}^2_+ & \text{if } y_2 = 0, \\ \text{cone conv } \{(y_1, |y_2|), (1, 0)\} & \text{otherwise.} \end{cases}$$

We use the sufficient criteria given in Theorem 5.11(a) to show that $\bar{y} = (0, 0)$ is a nondominated element of A w.r.t. $\mathcal{D}$. We choose $r = (1, 1)$, compare Fig. 5.4.

It holds $r \in \mathcal{D}(y)$ for all $y \in A$. Of course $\chi_{\bar{y}, r}(\bar{y}) = 0$.

For $y = (y_1, y_2) \in A \setminus \{\bar{y}\}$, $\mathcal{D}(y) \subset \mathbb{R}^2_+$ and $y_2 \geq y_1 > 0$ and thus

$$\begin{aligned}
\chi_{\bar{y}, r}(y) &= \inf\{t \in \mathbb{R} \mid (t, t) \in \{(y_1, y_2)\} + \mathcal{D}(y_1, y_2)\} \\
&\geq \inf\{t \in \mathbb{R} \mid (t, t) \in \{(y_1, y_2)\} + \mathbb{R}^2_+\} \\
&= \inf\{t \in \mathbb{R} \mid t \geq y_1, \ t \geq y_2\} \\
&= y_2 > 0 = \chi_{\bar{y}, r}(\bar{y}).
\end{aligned}$$

Note that the presented results cannot be applied if

$$\bigcap_{y \in A} \mathcal{D}(y) = \{0_Y\}.$$

5.2.2   *Characterization of Minimal Elements*

In this subsection, we give necessary and sufficient conditions for weakly minimal and minimal elements of some set A w.r.t. the ordering map $\mathcal{D}$ based on the functional introduced in (5.5).

Theorem 5.13. *Let $\bar{y} \in A$.*

(a) If $\bar{y}$ is a minimal element of A w.r.t. the ordering map $\mathcal{D}$, then for any pointed convex cone $K \subset \mathcal{D}(\bar{y})$ and any $r \in K \setminus \{0_Y\}$ the element $\bar{y}$ is a unique minimal solution of $\min_{y \in A} \psi_{\bar{y},r}(y)$, i.e.

$$\psi_{\bar{y},r}(y) > \psi_{\bar{y},r}(\bar{y}) = 0 \text{ for all } y \in A \setminus \{\bar{y}\}.$$

(b) If for any $a \in Y$, $r \in Y$ and any convex cone K with $\mathcal{D}(\bar{y}) \subset K$ the element $\bar{y}$ is a unique minimal solution of $\min_{y \in A} \psi_{a,r}(y)$, i.e.

$$\psi_{a,r}(y) > \psi_{a,r}(\bar{y}) \text{ for all } y \in A \setminus \{\bar{y}\},$$

then $\bar{y}$ is a minimal element of A w.r.t. the ordering map $\mathcal{D}$.

(c) Let $int(\mathcal{D}(\bar{y})) \neq \emptyset$. If $\bar{y}$ is a weakly minimal element of A w.r.t. the ordering map $\mathcal{D}$, then for any pointed convex cone $K \subset \mathcal{D}(\bar{y})$ with $int(K) \neq \emptyset$ and $r \in int(K)$ the element $\bar{y}$ is a minimal solution of $\min_{y \in A} \psi_{\bar{y},r}(y)$, i.e.

$$\psi_{\bar{y},r}(y) \geq \psi_{\bar{y},r}(\bar{y}) = 0 \text{ for all } y \in A \setminus \{\bar{y}\}.$$

(d) Let $int(\mathcal{D}(\bar{y})) \neq \emptyset$. If for any $a \in Y$, $r \in Y$ and any convex cone K with $\mathcal{D}(\bar{y}) \subset K$ the element $\bar{y}$ is a minimal solution of $\min_{y \in A} \psi_{a,r}(y)$, i.e.

$$\psi_{a,r}(y) \geq \psi_{a,r}(\bar{y}) \text{ for all } y \in A,$$

then $\bar{y}$ is a weakly minimal element of A w.r.t. the ordering map $\mathcal{D}$.

Proof. $\psi_{\bar{y},r}(\bar{y}) = \inf\{t \in \mathbb{R} \mid t\,r \in K\}$ and because of $r \in K \setminus \{0_Y\}$ and as K is a pointed convex cone this implies $\psi_{\bar{y},r}(\bar{y}) = 0$.

(a) If $\bar{y}$ is not a unique minimal solution of $\min_{y \in A} \psi_{\bar{y},r}(y)$, then there exists some $t \in \mathbb{R}$, $t \leq 0$, and some $y \in A \setminus \{\bar{y}\}$ such that $\bar{y} + t\,r - y \in K$. Because of $t\,r \in -K$ and as K is a convex cone, this implies

$$y \in \{\bar{y}\} - K \setminus \{0_Y\} \subset \{\bar{y}\} - \mathcal{D}(\bar{y}) \setminus \{0_Y\},$$

in contradiction to $\bar{y}$ a minimal element of A w.r.t. $\mathcal{D}$.

(b) For $\bar{t} := \psi_{a,r}(\bar{y})$ it holds

$$a + \bar{t}\,r - \bar{y} \in K.$$

If there is some $y \in A \setminus \{\bar{y}\}$ with $\bar{y} - y \in \mathcal{D}(\bar{y})$, then

$$a + \bar{t}\,r - y = a + \bar{t}\,r - \bar{y} + (\bar{y} - y) \in K + \mathcal{D}(\bar{y}) \subset K$$

and y is also a minimal solution of $\min_{z \in A} \psi_{a,r}(z)$, which is a contradiction.

(c) If there exists some $t < 0$ and some $y \in A$ such that $\bar{y} + t\,r - y \in K$, then this implies because of $t\,r \in -\text{int}(K)$ and by Lemma 1.7(iv)

$$y \in \{\bar{y}\} - \text{int}(K) \subset \{\bar{y}\} - \text{int}(\mathcal{D}(\bar{y})),$$

in contradiction to $\bar{y}$ a weakly minimal element of A w.r.t. $\mathcal{D}$.

(d) For $\bar{t} := \psi_{a,r}(\bar{y})$ it holds $a + \bar{t}\,r - \bar{y} \in K$. If there is some $y \in A$ with $\bar{y} - y \in \text{int}(\mathcal{D}(\bar{y}))$, then by Lemma 1.7(iv)

$$a + \bar{t}\,r - y = a + \bar{t}\,r - \bar{y} + (\bar{y} - y) \in K + \text{int}(\mathcal{D}(\bar{y})) \subset \text{int}K.$$

Then there exists some $\varepsilon > 0$ such that also $a + (\bar{t} - \varepsilon)\,r - y \in K$, i.e. $\psi_{a,r}(y) \le \bar{t} - \varepsilon$ in contradiction to $\bar{t} = \psi_{a,r}(\bar{y}) > \bar{t} - \varepsilon$.

$\square$

Summarizing the previous results we get for weakly minimal and minimal elements of some set A w.r.t. $\mathcal{D}$ the following complete characterization:

Corollary 5.14. *Let $\bar{y} \in A$.*

(a) *Set $K := \mathcal{D}(\bar{y})$ and choose $r \in \mathcal{D}(\bar{y}) \setminus \{0_Y\}$. $\bar{y}$ is a minimal element of A w.r.t. the ordering map $\mathcal{D}$ if and only if $\bar{y}$ is a unique minimal solution of $\min_{y \in A} \psi_{\bar{y},r}(y)$, i.e.*

$$\psi_{\bar{y},r}(y) > \psi_{\bar{y},r}(\bar{y}) = 0 \text{ for all } y \in A \setminus \{\bar{y}\}.$$

(b) *Let $\text{int}(\mathcal{D}(\bar{y})) \ne \emptyset$. Set $K := \mathcal{D}(\bar{y})$ and choose $r \in \text{int}(\mathcal{D}(\bar{y}))$. $\bar{y}$ is a weakly minimal element of A w.r.t. $\mathcal{D}$ if and only if $\bar{y}$ is a minimal solution of $\min_{y \in A} \psi_{\bar{y},r}(y)$, i.e.*

$$\psi_{\bar{y},r}(y) \ge \psi_{\bar{y},r}(\bar{y}) = 0 \text{ for all } y \in A.$$

Similar results as in Corollary 5.14 are obtained in [155, Theorem 3.1]. The result of Corollary 5.14 is also a direct consequence of Lemma 2.15 and known scalarization results for efficient elements, see for instance [128, 151].

Example 5.15. Let Y, $\mathcal{D}$ and A be specified as in Example 5.12, see also Example 2.14. We use Corollary 5.14(a) and choose $\bar{y} \in A \setminus \{(0,0)\}$ arbitrarily, $K = \mathcal{D}(\bar{y})$ and $r := \bar{y} \in \mathcal{D}(\bar{y}) \setminus \{(0,0)\}$. Then

$$\begin{aligned}
\psi_{\bar{y},r}(0,0) &= \inf\{t \in \mathbb{R} \mid \bar{y} + t\,r \in \mathcal{D}(\bar{y})\} \\
&= \inf\{t \in \mathbb{R} \mid (1+t)\,\bar{y} \in \mathcal{D}(\bar{y})\} \\
&= -1 < 0 = \psi_{\bar{y},r}(\bar{y}).
\end{aligned}$$

Thus, all elements in $A \setminus \{(0,0)\}$ are not minimal elements of A w.r.t. $\mathcal{D}$.

For $\hat{y} = (0,0)$ it holds for $K = \mathcal{D}(\hat{y}) = \mathbb{R}_+^2$ and $r = (1,1) \in K$

$$
\begin{aligned}
\psi_{\hat{y},r}(y) &= \inf\{t \in \mathbb{R} \mid (t,t) \in \{(y_1, y_2)\} + \mathbb{R}_+^2\} \\
&= \inf\{t \in \mathbb{R} \mid t \geq y_1, \ t \geq y_2\} \\
&= y_2 > 0 = \psi_{\hat{y},r}(\hat{y}) \qquad \text{for all } y \in A \setminus \{\hat{y}\}.
\end{aligned}
$$

Hence, $\hat{y} = (0,0)$ is the unique minimal element of A w.r.t. $\mathcal{D}$.

5.3 Notes on the Literature

The signed distance function Δ_S defined in (5.1) was introduced by Hiriart-Urruty in 1979 in [84, 85] to analyze the geometry of non-smooth optimization problems. The properties of the function were studied, among others, by Hiriart-Urruty [84], Zaffaroni [160], and Liu et al. [110]. Examples 5.2 and 5.3 are due to Zaffaroni [160]. The function was used in vector optimization in partially ordered spaces for characterizing efficient and weakly efficient elements by Zaffaroni [160]; and by Liu et al. [110] as a so-called merit function. For characterizing various optimal elements Ha used it in [77], see also [78]. She also provided the subdifferential of $\Delta_S(\cdot - \bar{y})$ for various sets S at the minimizer $\bar{y}$. The scalarization results of Sect. 5.1 for variable ordering structures using also a modification of the signed distance function were first presented in short form in [47]. The results are also discussed by Gebhardt in her diploma thesis [61] in which she also provided several examples illustrating the scalarization results.

The translative function defined in (5.5) was used as nonlinear separation and scalarization functional by Gerstewitz (Tammer) [64] in 1983, see also [65] by Gerstewitz (Tammer) and Iwanow, and [66] by Gerth (Tammer) and Weidner. It was also used earlier for instance in [104] by Krasnosel'skij to establish conditions for a cone to be normal. For an exhaustive review also of early references on this functional we refer to the literature survey given by Hamel in [80].

For characterizing solutions of vector optimization problems in partially ordered spaces the functional was for instance studied by Pascoletti and Serafini [128] or, for conic closed sets K, by Rubinov and Gasimov [133]. Later, Eichfelder [41–45] and Gourion and Luc [72] used it for a numerical solution method for vector optimization problems in partially ordered spaces. The functional can also be used to characterize properly efficient elements in a partially ordered space, see for instance Gerth (Tammer) and Weidner [66] or, for the notions of (generalized) Henig properly efficient elements, Qiu and Hao [131]. In the context of approximate solutions it was for instance used by Gutiérrez et al. in [76].

Its properties are well studied, see among others Weidner [151–153], Gerth (Tammer) and Weidner [66], Helbig [81, 82], Sterna-Karwat [138, 139], Chen et al. [31] and Tammer and Zălinescu [143]. We refer also to the book by Göpfert et al. [71] and for a literature survey about the usage of this functional and a study of the local Lipschitz properties of the functional under as weak as possible assumptions to Tammer and Zălinescu [142].

In the context of vector variational inequalities with variable ordering structures, Al-Homodan, Ansari and Schaible apply the translative functional as a so-called gap function in [3], cf. Sect. 10.1. Further, setting $K := \mathcal{D}(\bar{y})$, $r := r(\bar{y}) \in \text{int}(\mathcal{D}(\bar{y}))$ and $a := \bar{y}$ for some given $\bar{y} \in Y$, the functional $\psi_{a,r}$ in (5.5) was already examined in the context of variable ordering structures by Xiao et al. [155] in view of the binary relation $\leq_2$ defined in (1.2) which corresponds to the notion of minimal elements. The scalarization results in the way they are collected in Sect. 5.2 were first presented in short form in [47]. Soleimani and Tammer use the functional in [140] for characterizing approximate optimal solutions w.r.t. a variable ordering structure.

In addition to that, a modification of the scalarization $\psi_{a,r}$ was studied already by Chen and Yang [30] and later also by Chen and colleagues [29, 31]: for some $r \in \text{int}\left(\bigcap_{y \in Y} \mathcal{D}(y)\right)$ the functional defined by

$$(y, z) \in Y \times Y \mapsto \inf\{t \in \mathbb{R} \mid t\, r - y \in \mathcal{D}(z)\} \tag{5.10}$$

was discussed. Note that in (5.10) the cone-valued map $\mathcal{D}$ is evaluated in some element z independently of the choice of y while for determining $\chi_{a,r}(y)$ the cone $\mathcal{D}(y)$ is used. Under some assumptions including the linearity of the cone-valued map $\mathcal{D}$, see Definition 3.16, it was established that the function defined in (5.10) is convex. However, according to Lemma 3.17, this linearity condition implies that $\mathcal{D}$ is a trivial cone-valued map. In [31], Chen, Yang and Yu relax the assumption $r \in \text{int}\left(\bigcap_{y \in Y} \mathcal{D}(y)\right)$ by replacing r in (5.10) with $r(y)$ where $r(y) \in \text{int}(\mathcal{D}(y))$ for all $y \in Y$. They show that if $\mathcal{D}$ is upper semicontinuous then the functional in (5.10) is lower semicontinuous. Upper semicontinuity is according to Lemma 3.28 a strong assumption.

Another modification was studied by Allende and Tammer in [5]:

$$(y, r) \in Y \times Y \mapsto \inf\{t \in \mathbb{R} \mid y + t\, r \in A + \mathcal{D}(y)\} .$$

Characterizations of minimal elements of the set A w.r.t. $\mathcal{D}$ were given. For instance, if $\bar{y}$ is a minimal element of A w.r.t. $\mathcal{D}$ then for all $r \in \mathcal{D}(\bar{y}) \setminus \{0_Y\}$ it holds that $\inf\{t \in \mathbb{R} \mid \bar{y} + t\, r \in A + \mathcal{D}(\bar{y})\} = 0$. Also more general set-valued maps $\mathcal{D} : Y \to 2^Y$ defining some ordering structure were studied there.

Chapter 6
Scalarizations for Variable Orderings Given by Bishop-Phelps Cones

In this chapter, we restrict ourselves to the case when the values of the cone-valued map defining the variable ordering structure are Bishop-Phelps cones (BP cones). Doing this allows to formulate a nonlinear scalarization functional which completely characterizes nondominated (and also minimal) elements and for which convexity can be shown under some assumptions. With this functional at hand we convert the vector optimization problem into an optimization problem with a scalar-valued objective function, and obtain necessary and sufficient optimality conditions for the vector optimization problem in form of Fermat and Lagrange multiplier rule, see Chap. 7. Further, we give an existence result for (weakly) nondominated elements. We also present a scalarization functional which enables a complete characterization of strongly nondominated elements. Our main concern in this chapter are the different types of nondominated elements of a vector optimization problem with a variable ordering structure but for completeness we also present the corresponding results for minimal elements.

Recall that it is quite natural to restrict the examinations to BP cones because many cones such as the Lorentz cone and its extensions using various l_p norms in any finite dimensional space and the nonnegative orthants in the classical Banach spaces $L^1_{[0,1]}$, l^1 are BP cones, compare Sect. 1.2. Further, all closed convex cones with a bounded base are representable as a BP cone, see Theorem 1.17.

Thus, in this chapter we assume $(Y, \| \cdot \|)$ to be a real normed space and the cone-valued map $\mathcal{D}: Y \to 2^Y$ to be given by

$$\mathcal{D}(y) := C(\ell(y)) := \{u \in Y \mid \|u\|_y \le \ell(y)(u)\} \text{ for all } y \in Y$$

with $\ell : Y \to Y^*$ a given map and with $\| \cdot \|_y$ a norm in Y for each $y \in Y$, equivalent to $\| \cdot \|$, which depends on y. Let A be a nonempty subset of Y.

G. Eichfelder, *Variable Ordering Structures in Vector Optimization*, Vector Optimization, DOI 10.1007/978-3-642-54283-1_6, © Springer-Verlag Berlin Heidelberg 2014

6.1   Characterization of Nondominated Elements

In this section, we obtain scalarization results for (strongly, weakly) nondominated elements w.r.t. variable ordering structures defined by BP cones. This delivers the base for the Fermat rule and Lagrange multiplier rule for vector optimization problems with a variable ordering structure, which we present in Chap. 7.

To some given $\bar{y} \in Y$ we define the functionals $\theta: Y \to \mathbb{R}$, $\eta_{\bar{y}}: Y \to \mathbb{R}$, $\gamma_{\bar{y}}: Y \to \mathbb{R}$ and $\xi_{\bar{y}}: Y \to \mathbb{R}$ by

$$
\begin{aligned}
\theta(y) &:= \ell(y)(y) & \text{for all } y \in Y, \\
\eta_{\bar{y}}(y) &:= \ell(y)(y - \bar{y}) & \text{for all } y \in Y, \\
\gamma_{\bar{y}}(y) &:= \ell(y)(y - \bar{y}) - \|y - \bar{y}\|_y & \text{for all } y \in Y, \\
\xi_{\bar{y}}(y) &:= \ell(y)(y - \bar{y}) + \|y - \bar{y}\|_y & \text{for all } y \in Y.
\end{aligned}
\tag{6.1}
$$

Using the functionals (6.1) we can characterize a (weakly, strongly) nondominated element of a set A w.r.t. $\mathcal{D}$. Note that neither additional assumptions on $\mathcal{D}$—besides that the images of $\mathcal{D}$ are BP cones—nor convexity assumptions on the set A are presumed.

Theorem 6.1. *Let $\bar{y} \in A$.*

(a) $\bar{y}$ is a strongly nondominated element of A w.r.t. the ordering map $\mathcal{D}$ if and only if $\bar{y}$ is a minimal solution of $\min_{y \in A} \gamma_{\bar{y}}(y)$, i.e. if and only if

$$\gamma_{\bar{y}}(y) \geq \gamma_{\bar{y}}(\bar{y}) = 0 \text{ for all } y \in A.$$

(b) $\bar{y}$ is a nondominated element of A w.r.t. the ordering map $\mathcal{D}$ if and only if $\bar{y}$ is a unique minimal solution of $\min_{y \in A} \xi_{\bar{y}}(y)$, i.e. if and only if

$$\xi_{\bar{y}}(y) > \xi_{\bar{y}}(\bar{y}) = 0 \text{ for all } y \in A \setminus \{\bar{y}\}.$$

(c) Let $\|\ell(y)\|_{y,*} > 1$ (and hence, int $\mathcal{D}(y) \neq \emptyset$) for all $y \in A$. $\bar{y}$ is a weakly nondominated element of A w.r.t. the ordering map $\mathcal{D}$ if and only if $\bar{y}$ is a minimal solution of $\min_{y \in A} \xi_{\bar{y}}(y)$, i.e. if and only if

$$\xi_{\bar{y}}(y) \geq \xi_{\bar{y}}(\bar{y}) = 0 \text{ for all } y \in A.$$

Here, $\| \cdot \|_{y,*}$ denotes the dual norm of $\| \cdot \|_y$.

Proof.

(a) $\bar{y}$ is strongly nondominated of A w.r.t. the ordering map $\mathcal{D}$ if and only if

$$
\begin{aligned}
&y - \bar{y} \in \mathcal{D}(y) = \{u \in Y \mid \|u\|_y \leq \ell(y)(u)\} \text{ for all } y \in A \\
\Leftrightarrow\ & \ell(y)(y - \bar{y}) - \|y - \bar{y}\|_y \geq 0 \text{ for all } y \in A \\
\Leftrightarrow\ & \gamma_{\bar{y}}(y) \geq \gamma_{\bar{y}}(\bar{y}) = 0 \text{ for all } y \in A.
\end{aligned}
$$

(b) $\bar{y}$ is a nondominated element of A w.r.t. the ordering map $\mathcal{D}$ if and only if

$$\bar{y} - y \notin \mathcal{D}(y) \text{ for all } y \in A \setminus \{\bar{y}\}$$
$$\Leftrightarrow \|y - \bar{y}\|_y > \ell(y)(\bar{y} - y) \text{ for all } y \in A \setminus \{\bar{y}\}$$
$$\Leftrightarrow \ell(y)(y - \bar{y}) + \|y - \bar{y}\|_y > 0 \text{ for all } y \in A \setminus \{\bar{y}\}$$
$$\Leftrightarrow \xi_{\bar{y}}(y) > \xi_{\bar{y}}(\bar{y}) = 0 \text{ for all } y \in A \setminus \{\bar{y}\}.$$

(c) $\bar{y}$ is a weakly nondominated element of A w.r.t. the ordering map $\mathcal{D}$ if and only if $\bar{y} - y \notin \text{int}\mathcal{D}(y)$ for all $y \in A$. According to Lemma 1.16(iv) we have $\text{int}\mathcal{D}(y) = \{u \in Y \mid \|u\|_y < \ell(y)(u)\} \neq \emptyset$ and hence

$$\bar{y} - y \notin \text{int}\mathcal{D}(y) \text{ for all } y \in A$$
$$\Leftrightarrow \ell(y)(y - \bar{y}) + \|y - \bar{y}\|_y \geq 0 \text{ for all } y \in A$$
$$\Leftrightarrow \xi_{\bar{y}}(y) \geq \xi_{\bar{y}}(\bar{y}) = 0 \text{ for all } y \in A.$$

$\square$

Remark 6.2. (i) According to Lemma 2.23(i), any strongly nondominated element is nondominated and the "only if" part of the assertion (b) in the above theorem also is necessary for $\bar{y}$ to be a strongly nondominated element of A w.r.t. $\mathcal{D}$.

(ii) Additionally, the "if" part of the assertion (a) in the above theorem is also sufficient for $\bar{y}$ to be a nondominated element of A w.r.t. $\mathcal{D}$.

(iii) Assuming only $\text{int}(\mathcal{D}(y)) \neq \emptyset$ (instead of $\|\ell(y)\|_{y,*} > 1$, which is a sufficient condition for that) in the assertion (c) in the above theorem, we obtain due to Lemma 1.16(iv) that

$$\xi_{\bar{y}}(y) \geq \xi_{\bar{y}}(\bar{y}) = 0 \text{ for all } y \in A$$

is a necessary condition for $\bar{y}$ a weakly nondominated element w.r.t. the ordering map $\mathcal{D}$.

We illustrate the results of the above theorem with an example:

Example 6.3. Let Y be the Euclidean space $\mathbb{R}^2$ and the cone-valued map $\mathcal{D} \colon \mathbb{R}^2 \to 2^{\mathbb{R}^2}$ be defined by $\mathcal{D}(y) := C(\ell(y))$ with $\ell \colon \mathbb{R}^2 \to \mathbb{R}^2$ as in (1.15), i.e.

$$\ell(y_1, y_2) := \left(\frac{3 + \sin y_1}{2}, \frac{3 + \cos y_2}{2} \right)^{\mathsf{T}} \in [1, 2] \times [1, 2],$$

and with $\|\cdot\|_y := \|\cdot\|_2$ for all $y \in \mathbb{R}^2$, compare Example 1.28. Further, let

$$A := \{(y_1, y_2) \in \mathbb{R}^2 \mid y_1 \geq 0, \ y_2 \geq 0, \ y_2 \geq \pi - y_1\}.$$

Then $\mathbb{R}^2_+ \subset \mathcal{D}(y) \subset \text{cone conv}\{y^A, y^B\}$ with $y^A := (0.8, -0.6)$, $y^B := (-0.6, 0.8)$ for all $y \in \mathbb{R}^2$ and thus one can easily verify that $\bar{y} = (0, \pi)$ is a nondominated element of A w.r.t. the ordering map $\mathcal{D}$. This can also be shown using Theorem 6.1(b), because it obviously holds for all $y \in A \setminus \{\bar{y}\}$ with $y_2 \geq \pi$

$$\xi_{\bar{y}}(y) = \tfrac{3+\sin y_1}{2}(y_1 - 0) + \tfrac{3+\cos y_2}{2}(y_2 - \pi) + \|y - (0, \pi)^\top\|_2$$
$$> 0 = \xi_{\bar{y}}(\bar{y}).$$

For $y \in A \setminus \{\bar{y}\}$ with $y_2 < \pi$ we have $0 > y_2 - \pi \geq -y_1$ and thus

$$\xi_{\bar{y}}(y) \geq \tfrac{3+\sin y_1}{2} y_1 + \tfrac{3+\cos y_2}{2}(-y_1) + \sqrt{y_1^2 + (y_2 - \pi)^2}$$

$$> \underbrace{\left(\frac{\sin y_1 - \cos y_2}{2} \right)}_{\in[-1,1]} y_1 + y_1 \geq 0.$$

However, $\bar{y}$ it is not a strongly nondominated element of A w.r.t. $\mathcal{D}$ because of

$$\gamma_{\bar{y}}(3\pi/2, 0) = (1, 2)(3\pi/2, -\pi)^\top - \|(3\pi/2, -\pi)^\top\|_2 < 0 = \gamma_{\bar{y}}(\bar{y}).$$

In the remaining of this section we assume $\| \cdot \|_y := \| \cdot \|$ for all $y \in Y$. As noted in Sect. 1.2.3, the values of the map $\mathcal{D}$ reduce to the BP cones

$$\mathcal{D}(y) = C(\ell(y)) = \{u \in Y \mid \|u\| \leq \ell(y)(u)\},$$

compare (1.14), and the functionals (6.1) become

$$
\begin{aligned}
\theta(y) &= \ell(y)(y) & &\text{for all } y \in Y, \\
\eta_{\bar{y}}(y) &= \ell(y)(y - \bar{y}) & &\text{for all } y \in Y, \\
\gamma_{\bar{y}}(y) &= \ell(y)(y - \bar{y}) - \|y - \bar{y}\| & &\text{for all } y \in Y, \\
\xi_{\bar{y}}(y) &= \ell(y)(y - \bar{y}) + \|y - \bar{y}\| & &\text{for all } y \in Y.
\end{aligned}
\tag{6.2}
$$

Next, we study properties of the functionals (6.2). We show that these functionals inherit from ℓ such properties as continuity, lower semicontinuity and the Lipschitz property. We also study the convexity of θ, $\eta_{\bar{y}}$ and $\xi_{\bar{y}}$ which is of importance for using these scalarization functionals in the next chapter for deriving sufficient optimality conditions. We begin with showing that the functionals (6.2) inherit the continuity from ℓ.

Lemma 6.4. *Suppose that ℓ is continuous at $y \in Y$. Then θ, $\eta_{\bar{y}}$, $\gamma_{\bar{y}}$ and $\xi_{\bar{y}}$ also are continuous at y.*

Proof. First suppose $y = 0_Y$. From $|\ell(y')(y')| \leq \|\ell(y')\|_* \|y'\|$ we get

$$
\begin{aligned}
\lim_{y' \to 0_Y} |\theta(y') - \theta(0_Y)| &= \lim_{y' \to 0_Y} |\ell(y')(y') - \ell(0_Y)(0_Y)| \\
&= \lim_{y' \to 0_Y} |\ell(y')(y')| \\
&= 0,
\end{aligned}
$$

which means that θ is continuous at the origin.

Now suppose $y \neq 0_Y$. As ℓ is continuous at y, for any $\varepsilon > 0$ there exists $\delta > 0$ such that

$$\delta \leq \min \left\{ \frac{\varepsilon}{2\|\ell(y)\|_*} \, , \, \|y\| \right\}$$

such that

$$\|\ell(y') - \ell(y)\|_* \leq \frac{\varepsilon}{4\|y\|} \quad \text{whenever} \quad \|y' - y\| \leq \delta.$$

Then, for $y' \in Y$ satisfying $\|y' - y\| \leq \delta$, we have

$$\|y'\| \leq \|y' - y\| + \|y\| \leq \delta + \|y\| \leq 2\|y\|$$

and therefore,

$$
\begin{aligned}
|\theta(y') - \theta(y)| &= |\ell(y')(y') - \ell(y)(y)| \\
&= |(\ell(y') - \ell(y))(y') + \ell(y)(y' - y)| \\
&\leq \|(\ell(y') - \ell(y))\|_* \, \|y'\| + \|\ell(y)\|_* \, \|y' - y\| \\
&\leq (\varepsilon/(4\|y\|))(2\|y\|) + \|\ell(y)\|_* \, \delta \\
&\leq \varepsilon/2 + \varepsilon/2 = \varepsilon.
\end{aligned}
$$

Thus, the functional θ is continuous at y. The continuity of the functionals $\eta_{\bar{y}}$, $\gamma_{\bar{y}}$ and $\xi_{\bar{y}}$ follows from the continuity of the functional θ and the norm. $\qquad \square$

In case $Y = \mathbb{R}^n$ we can also speak about the lower semicontinuity of the functionals (6.2):

Lemma 6.5. *Let $Y = \mathbb{R}^n$ and $\ell = (\ell_1, \ldots, \ell_n)$. Suppose that ℓ_i $(i = 1, \ldots, n)$ are lower semicontinuous at $y = (y_1, \ldots, y_n) \in \mathbb{R}^n_+$. Then the functionals θ, $\eta_{\bar{y}}$, $\gamma_{\bar{y}}$ and $\xi_{\bar{y}}$ also are lower semicontinuous at y.*

Proof. Recalling that $\ell(y)(y) = y_1 \ell_1(y) + \ldots + y_n \ell_n(y)$ for all $y \in \mathbb{R}^n$ one can derive from the lower semicontinuity of the functionals ℓ_i $(i = 1, \ldots, n)$ at the point $y \in \mathbb{R}^n_+$ the one of the functionals θ, $\eta_{\bar{y}}$, $\gamma_{\bar{y}}$ and $\xi_{\bar{y}}$. $\qquad \square$

Recall that a map $g : X \to Z$ between the real normed spaces $(X, \| \cdot \|_X)$ and $(Z, \| \cdot \|_Z)$ is *Lipschitz* (of rank L) on some set $S \subset X$ if

$$\|g(x) - g(x')\|_Z \leq L\|x - x'\|_X \quad \text{for all} \quad x, x' \in S$$

and g is *Lipschitz near* $\bar{x} \in X$ if it is Lipschitz on some closed ball centered at $\bar{x}$. We show that the functionals (6.2) inherit the Lipschitz property from ℓ.

Lemma 6.6. *Suppose that ℓ is Lipschitz near $y \in Y$. Then θ, $\eta_{\bar{y}}$, $\gamma_{\bar{y}}$, $\xi_{\bar{y}}$ also are Lipschitz near y.*

Proof. Suppose that ℓ is Lipschitz of rank L on the closed ball $B(y, \rho) := \{z \in Y \mid \|z - y\| \leq \rho\}$ with $\rho > 0$. Then for $y^1, y^2 \in B(y, \rho)$ we have

$$\|y^1\| \leq \|y\| + \|y^1 - y\| \leq \|y\| + \rho$$

and

$$
\begin{aligned}
\|\ell(y^2)\|_* &= \|\ell(y) + (\ell(y^2) - \ell(y))\|_* \\
&\leq \|\ell(y)\|_* + \|\ell(y^2) - \ell(y)\|_* \\
&\leq \|\ell(y)\|_* + L\|y^2 - y\| \\
&\leq \|\ell(y)\|_* + L\rho
\end{aligned}
$$

and thus

$$
\begin{aligned}
|\theta(y^1) - \theta(y^2)| &= |\ell(y^1)(y^1) - \ell(y^2)(y^2)| \\
&= |(\ell(y^1) - \ell(y^2))(y^1) + \ell(y^2)(y^1 - y^2)| \\
&\leq \|\ell(y^1) - \ell(y^2)\|_* \|y^1\| + \|\ell(y^2)\|_* \|y^1 - y^2\| \\
&\leq L\|y^1 - y^2\| \|y^1\| + \|\ell(y^2)\|_* \|y^1 - y^2\| \\
&= (L\|y^1\| + \|\ell(y^2)\|_*) \|y^1 - y^2\| \\
&\leq (L(\|y\| + \rho) + (\|\ell(y)\|_* + L\rho)) \|y^1 - y^2\| \\
&= (\|\ell(y)\|_* + 2L\rho + L\|y\|) \|y^1 - y^2\|.
\end{aligned}
$$

Thus, θ is Lipschitz near y. The Lipschitz property of the functionals $\eta_{\bar{y}}$, $\gamma_{\bar{y}}$ and $\xi_{\bar{y}}$ near y follows from the Lipschitz property of the functional θ and the norm near that element. $\square$

Below we prove a result on the convexity of the functional $\xi_{\bar{y}}$ that will be used to formulate conditions for the existence of (weakly) nondominated elements as well as sufficient optimality conditions for vector optimization problems. Recall that a map $\ell \colon Y \to Y^*$ is called *monotone*, if

$$(\ell(y^1) - \ell(y^2))(y^1 - y^2) \geq 0 \text{ for all } y^1, y^2 \in Y.$$

Lemma 6.7. *Suppose that ℓ is linear and monotone. Then the functionals θ, $\eta_{\bar{y}}$ and $\xi_{\bar{y}}$ are convex.*

Proof. Let $y^1, y^2 \in Y$ and $\lambda_1, \lambda_2 \in [0, 1]$ such that $\lambda_1 + \lambda_2 = 1$. We have

$$
\begin{aligned}
\theta(\lambda_1 y^1 + \lambda_2 y^2) &= \ell(\lambda_1 y^1 + \lambda_2 y^2)(\lambda_1 y^1 + \lambda_2 y^2) \\
&= \lambda_1 \ell(y^1)(\lambda_1 y^1 + \lambda_2 y^2) + \lambda_2 \ell(y^2)(\lambda_1 y^1 + \lambda_2 y^2) \\
&= \lambda_1 \ell(y^1)(y^1) + \lambda_1 \lambda_2 \ell(y^1)(y^2 - y^1) \\
&\quad + \lambda_2 \ell(y^2)(y^2) + \lambda_1 \lambda_2 \ell(y^2)(y^1 - y^2) \\
&= \lambda_1 \theta(y^1) + \lambda_2 \theta(y^2) - \lambda_1 \lambda_2 (\ell(y^1) - \ell(y^2))(y^1 - y^2) \\
&\leq \lambda_1 \theta(y^1) + \lambda_2 \theta(y^2).
\end{aligned}
$$

Thus, θ is convex. Because of that and the linearity of ℓ and $\ell(y)$ for all $y \in Y$, we obtain for $y^1, y^2 \in Y$, $\lambda_1, \lambda_2 \in [0, 1]$ with $\lambda_1 + \lambda_2 = 1$

$$\begin{aligned}
\eta_{\bar{y}}(\lambda_1 y^1 + \lambda_2 y^2) &= \theta(\lambda_1 y^1 + \lambda_2 y^2) - \ell(\lambda_1 y^1 + \lambda_2 y^2)(\bar{y}) \\
&\leq \lambda_1 \theta(y^1) + \lambda_2 \theta(y^2) - \lambda_1 \ell(y^1)(\bar{y}) - \lambda_2 \ell(y^2)(\bar{y}) \\
&= \lambda_1 \eta_{\bar{y}}(y^1) + \lambda_2 \eta_{\bar{y}}(y^2),
\end{aligned}$$

i.e. the convexity of the functional $\eta_{\bar{y}}$. The convexity of $\xi_{\bar{y}}$ follows from the convexity of $\eta_{\bar{y}}$ and the norm. □

Recall that by Lemma 1.16(vi) the set $\{z \in \mathcal{D}(y) \mid \ell(y)(z) = 1\}$ defines a base of $\mathcal{D}(y)$ for all $y \in Y$ and thus, if ℓ is linear and monotone, then $\mathcal{D}$ is baselinear and basemonotone by Definitions 3.18 and 3.39.

Example 6.8. Let $Y = R^m$, M be a real positive semidefinite $m \times m$ matrix and $\ell(y) := My$ for all $y \in Y$. Then ℓ is linear, monotone and according to Lemma 6.7, the functionals θ, $\eta_{\bar{y}}$ and $\xi_{\bar{y}}$ are convex.

Next we use the scalarization functionals for existence results for nondominated elements. Additional existence results for $(\mathcal{D}(A))_{Y*}^{\#} \neq \emptyset$ were already given in Theorem 4.8.

Theorem 6.9. *Suppose that $A \subset Y$ is a nonempty compact set and ℓ is Lipschitz on A of rank L satisfying $L\|y\| \leq 1$ for all $y \in A$.*

(a) *If $\|\ell(y)\|_* > 1$ for all $y \in A$ then A has a weakly nondominated element w.r.t. $\mathcal{D}$.*

(b) *If $L\|y\| < 1$ for all $y \in A$ then A has a nondominated element w.r.t. $\mathcal{D}$.*

Proof. Since ℓ is Lipschitz, it is also continuous and Lemma 6.4 yields that θ is also continuous. Therefore, θ attains its minimum on the compact set A at, say $\bar{y} \in A$, i.e.

$$\ell(y)(y) \geq \ell(\bar{y})(\bar{y}) \quad \text{for all } y \in A \setminus \{\bar{y}\}$$

which implies

$$\begin{aligned}
\xi_{\bar{y}}(y) &= \ell(y)(y - \bar{y}) + \|y - \bar{y}\| \\
&\geq \ell(\bar{y})(\bar{y}) - \ell(y)(\bar{y}) + \|y - \bar{y}\| \\
&\geq -\|\ell(\bar{y}) - \ell(y)\|_* \|\bar{y}\| + \|y - \bar{y}\| \\
&\geq -L\|\bar{y} - y\| \|\bar{y}\| + \|y - \bar{y}\| \\
&= (1 - L\|\bar{y}\|)\|y - \bar{y}\| \geq 0 \quad \text{for all } y \in A \setminus \{\bar{y}\},
\end{aligned}$$

i.e. $\xi_{\bar{y}}(y) \geq 0$ for all $y \in A \setminus \{\bar{y}\}$. This means according to Theorem 6.1(c) that $\bar{y}$ is a weakly nondominated element of A in the case (a). When $L\|y\| < 1$ for all $y \in A$ one gets $\xi_{\bar{y}}(y) > 0$ for all $y \in A \setminus \{\bar{y}\}$ and $\bar{y}$ is a nondominated element of A in case (b) by Theorem 6.1(b). □

We illustrate Theorem 6.9 by the following example.

Example 6.10. Let Y be the Euclidean space $\mathbb{R}^2$, let the cone-valued map $\mathcal{D}: \mathbb{R}^2 \to 2^{\mathbb{R}^2}$ be defined by $\mathcal{D}(y) := C(\ell(y))$ with $\ell : \mathbb{R}^2 \to \mathbb{R}^2$ as in (1.15), i.e.

$$\ell(y_1, y_2) = \left(\frac{3 + \sin y_1}{2}, \frac{3 + \cos y_2}{2} \right)^{\mathsf{T}} \in [1, 2] \times [1, 2]$$

and let

$$A := \left\{ (y_1, y_2) \in \mathbb{R}^2 \mid y_1 \geq 0, \ y_2 \geq 0, \ \|y\| \leq 2 \right\}.$$

Then ℓ is Lipschitz of rank $L = 1/2$ and it holds $L\|y\| \leq 1$ and $\|\ell(y)\|_* > 1$ for all $y \in A$. Thus, according to Theorem 6.9(a) a weakly nondominated element of A w.r.t. $\mathcal{D}$ exists.

Moreover, we can show that $\bar{y} := (0,0)$ is a nondominated element of A w.r.t. $\mathcal{D}$. Indeed, since

$$\xi_{\bar{y}}(y_1, y_2) = \underbrace{\left(\frac{3 + \sin y_1}{2} \right)}_{>0} y_1 + \underbrace{\left(\frac{3 + \cos y_2}{2} \right)}_{>0} y_2 + \|(y_1, y_2) - (0,0)\|$$
$$> 0 = \xi_{\bar{y}}(\bar{y}) \text{ for all } (y_1, y_2) \in A \setminus \{\bar{y}\},$$

Theorem 6.1(b) implies that $(0,0)$ is a nondominated element of A w.r.t. $\mathcal{D}$.

6.2 Characterization of Minimal Elements

In this section, we obtain scalarization results for (strongly, weakly) minimal elements w.r.t. variable ordering structures defined by BP cones. To some given $\bar{y} \in Y$ we define now the functionals $\bar{\gamma}_{\bar{y}}: Y \to \mathbb{R}$ and $\xi_{\bar{y}}: Y \to \mathbb{R}$ by

$$\begin{aligned} \bar{\gamma}_{\bar{y}}(y) &:= \ell(\bar{y})(y - \bar{y}) - \|y - \bar{y}\|_{\bar{y}} \text{ for all } y \in Y, \\ \xi_{\bar{y}}(y) &:= \ell(\bar{y})(y - \bar{y}) + \|y - \bar{y}\|_{\bar{y}} \text{ for all } y \in Y. \end{aligned} \tag{6.3}$$

The functionals are special cases of $\gamma_{\bar{y}}$ and $\xi_{\bar{y}}$ for $\ell(y) := \ell(\bar{y})$ for all $y \in Y$. Thus, they are continuous, Lipschitz near any $y \in Y$ and $\xi_{\bar{y}}$ is convex.

We obtain the following characterization results:

Theorem 6.11. *Let $\bar{y} \in A$.*

(a) *$\bar{y}$ is a strongly minimal element of A w.r.t. the ordering map $\mathcal{D}$ if and only if $\bar{y}$ is a minimal solution of $\min_{y \in A} \bar{\gamma}_{\bar{y}}(y)$, i.e. if and only if*

$$\bar{\gamma}_{\bar{y}}(y) \geq \bar{\gamma}_{\bar{y}}(\bar{y}) = 0 \text{ for all } y \in A.$$

(b) $\bar{y}$ *is a minimal element of* A *w.r.t. the ordering map* $\mathcal{D}$ *if and only if* $\bar{y}$ *is a unique minimal solution of* $\min_{y \in A} \bar{\xi}_{\bar{y}}(y)$, *i.e. if and only if*

$$\bar{\xi}_{\bar{y}}(y) > \bar{\xi}_{\bar{y}}(\bar{y}) = 0 \ \text{for all } y \in A \setminus \{\bar{y}\}.$$

(c) *Let* $\|\ell(y)\|_* > 1$ *(and hence, int* $\mathcal{D}(y) \neq \emptyset$*) for all* $y \in A$. $\bar{y}$ *is a weakly minimal element of* A *w.r.t. the ordering map* $\mathcal{D}$ *if and only if* $\bar{y}$ *is a minimal solution of* $\min_{y \in A} \bar{\xi}_{\bar{y}}(y)$, *i.e. if and only if*

$$\bar{\xi}_{\bar{y}}(y) \geq \bar{\xi}_{\bar{y}}(\bar{y}) = 0 \ \text{for all } y \in A.$$

Proof. The proof is similar to the proof of Theorem 6.1. □

Example 6.12. Let Y, $\mathcal{D}$ and the set A be specified as in Example 6.3. Then $\bar{y} = (0, \pi)$ is also a minimal element but not a strongly minimal element of A w.r.t. the ordering map $\mathcal{D}$ because it holds

$$\bar{\xi}_{\bar{y}}(y) = (3/2)y_1 + (y_2 - \pi) + \|y - (0, \pi)^\top\|_2 > 0 = \bar{\xi}_{\bar{y}}(\bar{y})$$

for all $y \in A \setminus \{(0, \pi)\}$, and

$$\bar{\gamma}_{\bar{y}}(\pi, 0) = (3/2)\pi + (-\pi) - \|(\pi, -\pi)\|_2 = \pi/2 - \sqrt{2}\pi < 0 = \bar{\gamma}_{\bar{y}}(\bar{y}).$$

For $\mathcal{D}(y) \equiv K$ for some pointed convex cone K the notions of nondominated and minimal elements reduce to the concept of efficiency in a linear space partially ordered by the convex cone K. The functionals introduced in the preceding sections deliver thus also a scalarization approach for characterizing efficient elements. We conclude this section by the following characterization of efficient, strongly efficient and weakly efficient elements in a real normed space Y partially ordered by some BP cone $K \subset Y$ which can easily be derived from Theorems 6.1 and 6.11.

Theorem 6.13. *Let the real normed space* $(Y, \| \cdot \|)$ *be partially ordered by a BP cone K given by*

$$K = \{y \in Y \mid \|y\| \leq \phi(y)\}, \tag{6.4}$$

where ϕ *is an arbitrary continuous linear functional from the dual space* Y^*. *Let* $\bar{y} \in A$. *Define the functionals* $\bar{\xi}_{\bar{y}}$ *and* $\tilde{\gamma}_{\bar{y}}$ *as follows:*

$$\tilde{\gamma}_{\bar{y}}(y) = \phi(y - \bar{y}) - \|y - \bar{y}\| \ \text{for all } y \in Y$$

and

$$\bar{\xi}_{\bar{y}}(y) = \phi(y - \bar{y}) + \|y - \bar{y}\| \ \text{for all } y \in Y.$$

(a) $\bar{y}$ is a strongly efficient element of A w.r.t. the ordering cone K if and only if $\bar{y}$ is a minimal solution of $\min_{y \in A} \tilde{\gamma}_{\bar{y}}(y)$, i.e. if and only if

$$\tilde{\gamma}_{\bar{y}}(y) \geq \tilde{\gamma}_{\bar{y}}(\bar{y}) = 0 \ \text{for all } y \in A.$$

(b) $\bar{y}$ is an efficient element of A w.r.t. the ordering cone K if and only if $\bar{y}$ is a unique minimal solution of $\min_{y \in A} \tilde{\xi}_{\bar{y}}(y)$, i.e. if and only if

$$\tilde{\xi}_{\bar{y}}(y) > \tilde{\xi}_{\bar{y}}(\bar{y}) = 0 \ \text{for all } y \in A \setminus \{\bar{y}\}.$$

(c) Supposing that $\|\phi\|_* > 1$ (and hence, $\mathrm{int} K \neq \emptyset$), $\bar{y}$ is a weakly efficient element of A w.r.t. the ordering cone K if and only if $\bar{y}$ is a minimal solution of $\min_{y \in A} \tilde{\xi}_{\bar{y}}(y)$, i.e. if and only if

$$\tilde{\xi}_{\bar{y}}(y) \geq \tilde{\xi}_{\bar{y}}(\bar{y}) = 0 \ \text{for all } y \in A.$$

6.3 Notes on the Literature

The scalarization functionals for (weakly, strongly) nondominated elements based on the representation of the variable ordering structure by BP cones as presented in this chapter were introduced by Eichfelder and Ha in [52]. That similar scalarization results can also be obtained for minimal elements w.r.t. a variable ordering structure was mentioned in the conclusions in [52] and explicitly written down later in [47, Sect. 4.4.2.3]. The special case for a partially ordered linear space and the notion of efficiency was formulated in [52, Theorem 3.7]. The scalarization functionals and characterization results together with examples for illustrating them were also extensively studied by Gebhardt in her diploma thesis [61].

If the images of $\mathcal{D}$ are nontrivial closed pointed convex cones and are not given as BP cones, but we can find some $\ell(y) \in Y^*$ for all $y \in Y$ such that

$$\mathcal{D}(y) \subset \{u \in Y \mid \|u\| \leq \ell(y)(u)\} = C(\ell(y)) \tag{6.5}$$

then we can use the results of Theorem 6.1 for formulating sufficient conditions for (weakly) nondominated elements and necessary conditions for strongly nondominated elements. Property (6.5) corresponds to that the cones $\mathcal{D}(y)$ are supernormal, see Definition 1.23. According to Lemma 1.26 any supernormal cone is also representable as a BP cone. By Lemma 1.25, for any supernormal cone $\mathcal{D}(y)$ there exists $(\phi(y), \alpha(y)) \in (\mathcal{D}(y))^{a*}$ with $\alpha(y) \neq 0$ and we obtain for these elements from the augmented dual cone also a scalarization functional by

$$y \mapsto \phi(y)(y - \bar{y}) + \alpha(y) \|y - \bar{y}\|.$$

By setting

$$\ell(y) := \frac{1}{\alpha(y)}\phi(y)$$

these functionals correspond to the functional $\xi_{\bar{y}}$. They have already been introduced by Kasimbeyli [100] in partially ordered spaces, i.e. for $\mathcal{D}(y) = K$ for all $y \in Y$ with K a closed pointed convex cone. Kasimbeyli characterizes properly efficient elements of a set in the senses of Henig or Benson using the functional $\tilde{\xi}_{\bar{y}}$. For a generalization to variable ordering structures see [54].

Further, the scalarization functional $\tilde{\gamma}_{\bar{y}}$ was already introduced by Bednarczuk and Przybyla in [13] as an order preserving functional for the partial ordering introduced by the related BP cone. It was shown that the minimal solutions of this functional deliver weakly efficient elements of a vector optimization problem.

Chapter 7
Optimality Conditions for Vector Optimization Problems

In this chapter we consider optimization problems with a set-valued objective map mapping in a real linear space which is equipped with a variable ordering structure, compare problem (VP) on page 42. We concentrate on (weakly) nondominated elements and (weakly) nondominated solutions w.r.t. an ordering map and formulate a Fermat rule for the unconstrained case and a Lagrange multiplier rule for the case with constraints when the variable ordering structure is given by an ordering map with images BP cones.

For (weakly) minimal solutions necessary and sufficient optimality conditions using for instance linear scalarizations can easily be obtained using classical results in partially ordered spaces. This is based on Lemma 2.15 which relates (weakly) minimal and (weakly) efficient elements. For instance, for a necessary condition, we obtain that any weakly minimal solution $(\bar{x}, \bar{y})$ of a vector optimization problem w.r.t. $\mathcal{D}$ is also a weakly efficient solution of the vector optimization problem with the objective space partially ordered by $K := \mathcal{D}(\bar{y})$. In that case optimality conditions are known in the literature, e.g. for the single-valued case see [94, Theorem 7.4 and Corollary 7.21].

Assume that X and Y are real topological linear spaces and Y is equipped with a variable ordering structure defined by a cone-valued map $\mathcal{D}: Y \to 2^Y$ with $\mathcal{D}(y)$ a pointed convex cone for all $y \in Y$. If we speak of an efficient element we additionally assume the space Y to be partially ordered by some pointed convex cone $K \subset Y$. Let $F: X \to 2^Y$ be a given set-valued map and $S \subset X$ a nonempty set. Recall that the vector optimization problem (VP) is defined as follows:

$$\text{Minimize} \quad F(x) \text{ subject to } x \in S. \tag{VP}$$

The various notions of nondominated (and minimal) elements w.r.t. the ordering map $\mathcal{D}$ for sets naturally induce corresponding notions of solutions to the optimization problem (VP) as given in Definition 2.33.

In the following we additionally assume that Y is a real normed space, $\ell: Y \to Y^*$ is a map and the ordering map $\mathcal{D}: Y \to 2^Y$ is given by

$$\mathcal{D}(y) = \{u \in Y \mid \|u\| \leq \ell(y)(u)\} \text{ for all } y \in Y.$$

G. Eichfelder, *Variable Ordering Structures in Vector Optimization*, Vector Optimization,   117
DOI 10.1007/978-3-642-54283-1_7, © Springer-Verlag Berlin Heidelberg 2014

Then we can use the scalarization functionals introduced in Sect. 6.1 in (6.1) for a scalar characterization of different kinds of locally nondominated solutions of the problem (VP). Recall that for some given $\bar{y} \in Y$ the functionals have been defined by

$$
\begin{aligned}
\theta(y) &= \ell(y)(y) \text{ for all } y \in Y, \\
\eta_{\bar{y}}(y) &= \ell(y)(y - \bar{y}) \text{ for all } y \in Y, \\
\gamma_{\bar{y}}(y) &= \ell(y)(y - \bar{y}) - \|y - \bar{y}\| \text{ for all } y \in Y, \\
\xi_{\bar{y}}(y) &= \ell(y)(y - \bar{y}) + \|y - \bar{y}\| \text{ for all } y \in Y.
\end{aligned}
\tag{7.1}
$$

Corollary 7.1. *Let $\bar{x} \in S$ and $\bar{y} \in F(\bar{x})$.*

(a) *$(\bar{x}, \bar{y})$ is a locally strongly nondominated solution of the problem (VP) w.r.t. the ordering map $\mathcal{D}$ if and only if there exists a neighborhood $U_{\bar{x}}$ of $\bar{x}$ such that the functional $\gamma_{\bar{y}}$ attains its minimum over $F(S \cap U_{\bar{x}})$ at $\bar{y}$, which means that*

$$
\gamma_{\bar{y}}(y) \geq \gamma_{\bar{y}}(\bar{y}) = 0, \text{ for all } y \in F(S \cap U_{\bar{x}}).
$$

(b) *$(\bar{x}, \bar{y})$ is a locally nondominated solution of the problem (VP) w.r.t. the ordering map $\mathcal{D}$ if and only if there exists a neighborhood $U_{\bar{x}}$ of $\bar{x}$ such that the functional $\xi_{\bar{y}}$ attains its strict minimum over $F(S \cap U_{\bar{x}})$ at $\bar{y}$, which means that*

$$
\xi_{\bar{y}}(y) > \xi_{\bar{y}}(\bar{y}) = 0 \text{ for all } y \in F(S \cap U_{\bar{x}}) \setminus \{\bar{y}\}.
$$

(c) *Supposing that $\|\ell(y)\|_* > 1$ (and hence int $\mathcal{D}(y) \neq \emptyset$) for all $y \in F(S)$, $(\bar{x}, \bar{y})$ is a locally weakly nondominated solution of the problem (VP) w.r.t. the ordering map $\mathcal{D}$ if and only if there exists a neighborhood $U_{\bar{x}}$ of $\bar{x}$ such that the functional $\xi_{\bar{y}}$ attains its minimum over $F(S \cap U_{\bar{x}})$ at $\bar{y}$, which means that*

$$
\xi_{\bar{y}}(y) \geq \xi_{\bar{y}}(\bar{y}) = 0 \text{ for all } y \in F(S \cap U_{\bar{x}}).
$$

(d) *Putting $U_{\bar{x}} = X$ in the assertions (a)–(c), we obtain necessary and sufficient conditions for corresponding global solutions.*

Proof. This result is a direct consequence of the definition of locally/globally (strongly, weakly) nondominated solutions of (VP) and Theorem 6.1.   □

Below we show that under some additional convexity assumptions, both the concepts of locally and globally weakly nondominated solutions of the problem (VP) w.r.t. the ordering map $\mathcal{D}$ coincide. Recall that the graph of F is the set

$$
\text{gr} F = \{(x, y) \in X \times Y \mid y \in F(x)\}.
$$

Lemma 7.2. *Suppose that $\|\ell(y)\|_* > 1$ (hence, int $\mathcal{D}(y) \neq \emptyset$) for all $y \in F(S)$. Let $\bar{x} \in S$ and $\bar{y} \in F(\bar{x})$. Assume that the graph of F is convex and $\xi_{\bar{y}}$ is convex. If $(\bar{x}, \bar{y})$ is a locally weakly nondominated solution of the problem (VP) w.r.t. the* *ordering map* $\mathcal{D}$ *then it is also a globally weakly nondominated solution of the problem* (VP) *w.r.t. the ordering map* $\mathcal{D}$.

Proof. As $(\bar{x}, \bar{y})$ is a locally weakly nondominated solution of the problem (VP) w.r.t. the ordering map $\mathcal{D}$, Corollary 7.1(c) implies that there exists some neighborhood $U_{\bar{x}}$ of $\bar{x}$ such that

$$\xi_{\bar{y}}(y) \geq \xi_{\bar{y}}(\bar{y}) = 0 \text{ for all } y \in F(S \cap U_{\bar{x}}). \tag{7.2}$$

By the definition of a globally weakly nondominated solutions of (VP) and Corollary 7.1(d), we have to show that

$$\xi_{\bar{y}}(y) \geq \xi_{\bar{y}}(\bar{y}) = 0 \text{ for all } y \in F(S). \tag{7.3}$$

Let $x \in S$ and $y \in F(x)$ be arbitrarily chosen. Since $(x, y), (\bar{x}, \bar{y}) \in \text{gr } F$ and the graph of F is convex, there exists some $\lambda \in (0, 1)$ such that

$$(x^\lambda, y^\lambda) := \lambda (x, y) + (1 - \lambda) (\bar{x}, \bar{y}) \in \text{gr } F$$

with $x^\lambda \in S \cap U_{\bar{x}}$ and $y^\lambda \in F(S \cap U_{\bar{x}})$. By (7.2), we have $\xi_{\bar{y}}(y^\lambda) \geq \xi_{\bar{y}}(\bar{y}) = 0$, which together with the convexity of $\xi_{\bar{y}}$ implies

$$0 \leq \xi_{\bar{y}}(y^\lambda) \leq \lambda \xi_{\bar{y}}(y) + (1 - \lambda)\xi_{\bar{y}}(\bar{y}) = \lambda \xi_{\bar{y}}(y).$$

Thus $\xi_{\bar{y}}(y) \geq 0$ for all $y \in F(S)$ which means that (7.3) holds. $\qquad \square$

If ℓ is linear and monotone, then Lemma 6.7 implies that $\xi_{\bar{y}}$ is convex and thus we get the following result.

Corollary 7.3. *Suppose that* $\|\ell(y)\|_* > 1$ *(hence,* $\text{int } \mathcal{D}(y) \neq \emptyset$*) for all* $y \in F(S)$. *Let* $\bar{x} \in S$ *and* $\bar{y} \in F(\bar{x})$. *Assume that the graph of* F *is convex and* ℓ *is linear and monotone. If* $(\bar{x}, \bar{y})$ *is a locally weakly nondominated solution of the problem* (VP) *w.r.t. the ordering map* $\mathcal{D}$ *then it is also a globally weakly nondominated solution of the problem* (VP) *w.r.t. the ordering map* $\mathcal{D}$.

7.1 Subdifferential and Coderivative

We recall the concepts of the *subdifferential* of a function and the *coderivative* of a set-valued map that will be used to formulate the optimality conditions. There are three types of normal cones and corresponding three types of subdifferentials and coderivatives which we sum up in the following [34, 56, 88, 117–120, 145].

Throughout this section, let X and Y be real normed spaces. We use the same notation $\| \cdot \|$ for the norms in X and Y and $\| \cdot \|_*$ for the induced norms in X^* and Y^*. The norm in $X \times Y$ is given by $\|(x, y)\| = \|x\| + \|y\|$. For a nonempty set $S \subset X$,

$$d(x; S) := d_S(x) = \inf\{\|x - s\| \mid s \in S\}$$

is again the *distance* from x to S and χ_S is the *indicator function* associated with S, i.e.,

$$\chi_S(x) := \begin{cases} 0 & \text{if } x \in S, \\ \infty & \text{otherwise.} \end{cases}$$

Assume that $g: X \to \mathbb{R} \cup \{\infty\}$ is a function. The *domain* and *epigraph* of g are the sets

$$\text{dom } g := \{x \in X \mid g \text{ is finite at } x\}$$

and

$$\text{epi } g := \{(x, t) \in X \times \mathbb{R} \mid g(x) \le t\},$$

respectively.

To define the concept of the subdifferential of g, suppose first that g is Lipschitz near $x \in \text{dom } g$. The *Ioffe approximate subdifferential* of g at x is the set

$$\partial_A g(x) := \bigcap_{L \in \mathcal{F}} \limsup_{(\varepsilon, y) \to (0^+, x)} \partial_\varepsilon^- g_{y+L}(y), \tag{7.4}$$

where $\mathcal{F}$ is the collection of all finite dimensional subspaces of X,

$$g_{y+L}(u) := \begin{cases} g(u) & \text{if } u \in \{y\} + L, \\ \infty & \text{otherwise,} \end{cases}$$

and for $\varepsilon \ge 0$

$$\partial_\varepsilon^- g_{y+L}(y) := \left\{ x^* \in X^* \,\middle|\, x^*(v) \le \varepsilon\|v\| \right.$$
$$\left. + \liminf_{t \to 0^+} \frac{g_{y+L}(y+tv) - g_{y+L}(y)}{t} \,\, \forall v \in X \right\}.$$

The *Clarke generalized subdifferential* of g at x is the set

$$\partial_C g(x) := \{x^* \in X^* \mid x^*(v) \le g^0(x; v) \text{ for all } v \in X\}, \tag{7.5}$$

where $g^0(x; v)$ is the *generalized directional derivative* of g at x in the direction v

$$g^0(x; v) := \limsup_{(t, y) \to (0^+, x)} \frac{g(y + tv) - g(y)}{t}.$$

Let Ω be a nonempty subset of X different from X and $x \in \mathrm{cl}\Omega$. The *Ioffe approximate normal cone* to Ω at $x \in \Omega$ is given by

$$N_A(x;\Omega) := \bigcup_{\lambda>0} \lambda \partial_A d(x;\Omega)$$

and the *Clarke normal cone* to Ω at $x \in \Omega$ by

$$N_C(x;\Omega) := \mathrm{cl}\left(\bigcup_{\lambda>0} \lambda \partial_C d(x;\Omega)\right).$$

The *Mordukhovich normal cone* to Ω at x is defined by

$$N_M(x;\Omega) := \limsup_{x' \xrightarrow{\Omega} x, \varepsilon \to 0+} \hat{N}_\varepsilon(x';\Omega),$$

where the limit in the right-hand side means the sequential Kuratowski-Painlevé upper limit with respect to the norm topology of X and the weak-star topology w^* of X^*, i.e. for a set-valued map $G: X \to 2^{X^*}$

$$\limsup_{x' \to x} G(x') := \{x^* \in X^* \mid \exists x_k \to x \text{ and } x_k^* \xrightarrow{w^*} x^* \text{ with}$$
$$x_k^* \in G(x_k) \text{ for all } k \in \mathbb{N}\}.$$

$x' \xrightarrow{\Omega} x$ refers to all sequences converging to x that remain in Ω and $\hat{N}_\varepsilon(x';\Omega)$ is the set of *Fréchet ε-normals* to Ω at x' given by

$$\hat{N}_\varepsilon(x';\Omega) := \left\{ x^* \in X^* \,\middle|\, \limsup_{x'' \xrightarrow{\Omega} x'} \frac{x^*(x''-x')}{\|x''-x'\|} \le \varepsilon \right\}.$$

When $\varepsilon = 0$ this set is a cone which is called the *Fréchet normal cone* to Ω at x' and is denoted by $\hat{N}(x';\Omega)$. For $x \in \Omega$, $N_M(x;\Omega) \subset N_A(x;\Omega)$ and if X is finite dimensional then $N_M(x;\Omega) = N_A(x;\Omega)$.

Let $F: X \to 2^Y$ be a set-valued map. For the sake of convenience we assume that $F(x)$ is nonempty for all $x \in X$. Assume that the function g is lower semicontinuous and the set-valued map F is closed, i.e., its graph is a closed set. The subdifferentials for g and coderivatives for F in the sense of Ioffe, Clarke or Mordukhovich are defined through the corresponding normal cone as follows: $\partial g: X \to 2^{X^*}$ with

$$\partial g(x) := \{x^* \in X^* \mid (x^*,-1) \in N((x,g(x)); \mathrm{epi}\, g)\} \tag{7.6}$$

and $D^*F(x, y): Y^* \to 2^{X^*}$ with

$$D^*F(x, y)(y^*) := \{x^* \in X^* \mid (x^*, -y^*) \in N((x, y); \operatorname{gr} F)\}.$$

For the sake of convenience, we make the convention that the same notations N, ∂g and D^*F are used within this book for the normal cones, subdifferentials and coderivatives in the above senses and that the spaces under consideration are Asplund, i.e., each of its separable subspaces has a separable dual, whenever the subdifferential and the coderivative are understood in the sense of Mordukhovich. Asplund spaces are a broad class of Banach spaces which include all reflexive spaces and all spaces with separable duals. The product of Asplund spaces is Asplund as well. Note that the subdifferential in the sense of Ioffe (respectively, of Clarke) as in (7.6) is a generalization of the one given in (7.4) (respectively, in (7.5)) from Lipschitz to lower semicontinuous functions and both definitions coincide in the Lipschitzian case.

Next we recall some properties of the above normal cones, subdifferentials and coderivatives that will be used in the sequel. We denote by $\mathcal{L}(X, Y)$ the space of continuous linear maps from X to Y. Let a map $h: X \to Y$ be given. The map h is said to admit a *strict derivative* at x, which is an element of $\mathcal{L}(X, Y)$ denoted by $h'(x)$, provided that for each v the following holds:

$$\lim_{(t, x') \to (0^+, x)} \frac{h(x' + tv) - h(x')}{t} = h'(x)(v),$$

and provided the convergence is uniform for v in compact sets (this condition automatically holds if h is Lipschitz near x).

We need the following characterization:

Lemma 7.4 ([34, Proposition 2.2.1]). *Let h map a neighborhood of $x \in X$ to Y and let $\zeta \in \mathcal{L}(X, Y)$. The following are equivalent:*

(i) h *is strictly differentiable at x and $h'(x) = \zeta$.*
(ii) h *is Lipschitz near x, and for each $v \in X$ one has*

$$\lim_{(t, x') \to (0^+, x)} \frac{h(x' + tv) - h(x')}{t} = \zeta(v).$$

The following lemma provides the relation of the above concepts to the well-known concepts of the subdifferential and the normal cone known from convex analysis as well as some basic properties as the sum rule or the subdifferential of the norm [34, 56, 88, 117–120].

Lemma 7.5. *Assume that $g : X \to \mathbb{R} \cup \{+\infty\}$ is lower semicontinuous and Ω is a nonempty closed subset of X.*

(i) *If the function g is strictly differentiable near $x \in$ dom g, then $\partial g(x) = \{g'(x)\}$.*

(ii) *If g is convex and Lipschitz near $x \in$ dom g, then the above subdifferentials reduce to the subdifferential of convex analysis, i.e.,*

$$\partial g(x) = \{x^* \in X^* \mid x^*(x' - x) \le g(x') - g(x) \text{ for all } x' \in \text{dom } g\}.$$

(iii) *If $g(x') \ge g(x)$ for all x' in a neighborhood of $x \in$ dom g, then $0_{X^*} \in \partial g(x)$.*

(iv) *(sum rule) Assume that $h : X \to \mathbb{R} \cup \{+\infty\}$ is Lipschitz near the element $x \in$ dom $g \cap$ dom h, then*

$$\partial(g + h)(x) \subset \partial g(x) + \partial h(x)$$

and equality holds if at least one function is strictly differentiable near x.

(v) *$N(x; \Omega) = \partial \chi_\Omega(x) = \mathbb{R}_+ \partial d(x; \Omega)$ and if Ω is convex, then the above normal cones reduce to the normal cone of convex analysis, i.e. to the set*

$$\{x^* \in X^* \mid x^*(x' - x) \le 0 \text{ for all } x' \in \Omega\}.$$

(vi) *For the norm $\|\cdot\|$ in X one has:*

$$\partial\|\cdot\|(0_X) = B_{X^*},$$
$$\partial(-\|\cdot\|)(0_X) = B_{X^*}$$

if the subdifferential is understood in the sense of Clarke and

$$\partial(-\|\cdot\|)(0_X) = S_{X^*}$$

if the subdifferential is understood in the senses of Ioffe or Mordukhovich. Here B_{X^*} and S_{X^*} are the closed unit ball and the unit sphere in X^*, respectively.

Lemma 7.6. *Assume that a map $g: X \to Y$ is strictly differentiable near $\bar{x} \in X$, then*

$$D^* g(\bar{x}, g(\bar{x})) = \{[g'(\bar{x})]^*\}.$$

Here, $[g'(\bar{x})]^*: Y^* \to X^*$ is the adjoint map to $g'(\bar{x})$ defined by

$$[g'(\bar{x})]^*(y^*)(x) = y^*(g'(\bar{x})(x))$$

for all $(x, y^*) \in X \times Y^*$.

Below we recall two results related to the Clarke penalization [34, Proposition 2.4.3] which will play an important role for obtaining the Fermat and Lagrange multiplier rules.

Lemma 7.7. *Let S be a subset of a Banach space X and $f : S \to \mathbb{R}$ be Lipschitz of rank L on S. Let $\bar{x}$ belong to a set $U \subset S$ and suppose that f attains a minimum over U at $\bar{x}$. Then for any $L' \geq L$, the function $\bar{f} : S \to \mathbb{R}$,*

$$\bar{f}(x) := f(x) + L' \, d(x; U) \text{ for all } x \in S$$

attains a minimum over S at $\bar{x}$. If $L' > L$ and U is closed, then any other point minimizing $\bar{f}$ over S also lies in U.

Corollary 7.8. *Let S be a subset of a Banach space X and suppose that $f : S \to \mathbb{R}$ is Lipschitz near $\bar{x} \in S$ and attains a minimum over S at $\bar{x}$. Then*

$$0_{X^*} \in \partial f(\bar{x}) + N(\bar{x}; S).$$

Note that the statement of Corollary 7.8 has been established in [34, Corollary, p. 52] for the subdifferential and normal cone in the sense of Clarke, but it remains true also for the subdifferential and normal cone in the senses of Ioffe and Mordukhovich [52].

For the normal cone in the product space $X \times Y$ it holds: let $\Omega_X \subset X$, $\Omega_Y \subset Y$ be nonempty sets unequal to the space and $(x, y) \in \text{cl}(\Omega_X) \times \text{cl}(\Omega_Y)$. Then

$$N((x, y); \Omega_X \times \Omega_Y) = N(x; \Omega_X) \times N(y; \Omega_Y).$$

We consider again the cone-valued map $\mathcal{D}$ defined by a map $\ell : Y \to Y^*$ and $\mathcal{D}(y) := \{u \in Y \mid \|u\| \leq \ell(y)(u)\}$ for all $y \in Y$ and the functionals in (7.1). Recall that by Lemma 6.6 the functionals inherit the Lipschitz property from ℓ. In the following lemma we also provide some formula for their derivatives and subdifferentials.

Lemma 7.9. *Suppose that ℓ is Lipschitz near $y \in Y$. Then*

$$\partial \gamma_{\bar{y}}(y) \subset \partial \eta_{\bar{y}}(y) + B_{Y^*} \text{ and } \partial \xi_{\bar{y}}(y) \subset \partial \eta_{\bar{y}}(y) + B_{Y^*}. \tag{7.7}$$

Moreover, if ℓ is Lipschitz near $\bar{y}$ then $\eta_{\bar{y}}$ is strictly differentiable at $\bar{y}$ and one has

$$\eta'_{\bar{y}}(\bar{y}) = \ell(\bar{y}) \tag{7.8}$$

and

$$\partial \gamma_{\bar{y}}(\bar{y}) \subset \{\ell(\bar{y})\} + B_{Y^*} \text{ and } \partial \xi_{\bar{y}}(\bar{y}) = \{\ell(\bar{y})\} + B_{Y^*}. \tag{7.9}$$

Proof. The Lipschitz property of the functionals $\eta_{\bar{y}}$, $\gamma_{\bar{y}}$ and $\xi_{\bar{y}}$ near y was shown in Lemma 6.6. It is easy to see that (7.7) follows from Lemma 7.5(iv) and (vi).

Now assume that ℓ is Lipschitz of rank L on the ball $B(\bar{y}, \rho)$ with $\rho > 0$. Our next aim is to prove that $\eta_{\bar{y}}$ is strictly differentiable at $\bar{y}$ and (7.8) holds. By Lemma 7.4, it suffices to show that for each $v \in Y$ one has

$$\lim_{(t,y')\to(0^+,\bar{y})} \frac{\eta_{\bar{y}}(y' + tv) - \eta_{\bar{y}}(y')}{t} = \ell(\bar{y})(v). \tag{7.10}$$

By the definition of the functional $\eta_{\bar{y}}$ it holds

$$\eta_{\bar{y}}(y' + tv) - \eta_{\bar{y}}(y') = \ell(y' + tv)(y' + tv - \bar{y}) - \ell(y')(y' - \bar{y}).$$

Further, as ℓ is Lipschitz near $\bar{y}$, one has for y' sufficiently close to $\bar{y}$ and t sufficiently small

$$\left| \frac{\eta_{\bar{y}}(y' + tv) - \eta_{\bar{y}}(y')}{t} - \ell(\bar{y})(v) \right|$$

$$= \left| \frac{\ell(y' + tv)(y' + tv - \bar{y}) - \ell(y')(y' - \bar{y})}{t} - \ell(\bar{y})(v) \right|$$

$$= \left| \frac{(\ell(y' + tv) - \ell(y'))(y' - \bar{y})}{t} + \frac{\ell(y' + tv)(tv)}{t} - \ell(\bar{y})(v) \right|$$

$$\leq \left| \frac{(\ell(y' + tv) - \ell(y'))(y' - \bar{y})}{t} \right| + \left| (\ell(y' + tv) - \ell(\bar{y}))(v) \right|$$

$$\leq \frac{L\|y' + tv - y'\| \, \|y' - \bar{y}\|}{t} + \left| (\ell(y' + tv) - \ell(\bar{y}))(v) \right|$$

$$\leq L\|v\| \, \|y' - \bar{y}\| + L \, \|y' + tv - \bar{y}\| \, \|v\|$$

which yields (7.10).

Finally, (7.9) follows from (7.8), Lemma 7.5(i) and (7.7).   □

Lemma 7.9 can be applied for instance to the case when ℓ is the map considered in Example 1.28.

7.2 Unconstrained Vector Optimization Problems

In this section we consider the unconstrained vector optimization problem

$$\text{Minimize } F(x) \text{ subject to } x \in X, \tag{UVP}$$

where $F: X \to 2^Y$ is a set-valued map. We further assume that X and Y are Banach spaces and the variable ordering structure is defined by a map $\mathcal{D}: Y \to 2^Y$ with $\mathcal{D}(y) = C(\ell(y))$ for all $y \in Y$, i.e.

$$\mathcal{D}(y) = \{u \in Y \mid \|u\| \leq \ell(y)(u)\} \text{ for all } y \in Y$$

for some given map $\ell: Y \to Y^*$.

The next theorem establishes necessary and sufficient conditions for the problem (UVP) in the form of the Fermat rule. Recall that a set-valued map F is called closed if the graph of F is closed.

Theorem 7.10. *Assume that F is closed, $\bar{x} \in X$ and $\bar{y} \in F(\bar{x})$ and ℓ is Lipschitz near $\bar{y}$.*

(a) *(Necessary condition) If $(\bar{x}, \bar{y})$ is a locally nondominated solution of (UVP) w.r.t. the ordering map $\mathcal{D}$ then*

$$0_{X^*} \in D^* F(\bar{x}, \bar{y})(y^*) \text{ for some } y^* \in \partial \xi_{\bar{y}}(\bar{y}) = \{\ell(\bar{y})\} + B_{Y^*}. \qquad (7.11)$$

(b) *(Necessary and sufficient conditions) Suppose that $\|\ell(y)\|_* > 1$ (hence, int $\mathcal{D}(y) \neq \emptyset$) for all $y \in F(X)$. Then (7.11) is necessary for $(\bar{x}, \bar{y})$ to be a locally weakly nondominated solution of (UVP) w.r.t. the ordering map $\mathcal{D}$.*

If we assume that the graph of F is convex and $\xi_{\bar{y}}$ is convex (in particular, ℓ is linear and monotone), then (7.11) also is sufficient for $(\bar{x}, \bar{y})$ to be a globally weakly nondominated solution of (UVP) w.r.t. the ordering map $\mathcal{D}$.

Proof. (a) First observe that the functional $\xi_{\bar{y}}$ is Lipschitz near $\bar{y}$ according to Lemma 6.6. Corollary 7.1(b) implies that there exists some neighborhood $U_{\bar{x}}$ of $\bar{x}$ such that the functional $\xi_{\bar{y}}$ attains its minimum over $F(U_{\bar{x}})$ at $\bar{y}$. We define a function $q : X \times Y \to \mathbb{R}$ by

$$q(x, y) := \xi_{\bar{y}}(y) \text{ for all } (x, y) \in X \times Y.$$

Let $U_{\bar{y}}$ be some neighborhood of $\bar{y}$. One can easily check that $(\bar{x}, \bar{y})$ is a minimizer of the functional q on $(U_{\bar{x}} \times U_{\bar{y}}) \cap \mathrm{gr} F$. Applying Corollary 7.8, we get

$$0_{X^* \times Y^*} \in \partial q(\bar{x}, \bar{y}) + N((\bar{x}, \bar{y}); (U_{\bar{x}} \times U_{\bar{y}}) \cap \mathrm{gr} F)$$

which yields

$$(0_{X^*}, 0_{Y^*}) \in (\{0_{X^*}\} \times \partial \xi_{\bar{y}}(\bar{y})) + N((\bar{x}, \bar{y}); \mathrm{gr} F).$$

Hence, there exists $y^* \in \partial \xi_{\bar{y}}(\bar{y})$ with $(0_{X^*}, -y^*) \in N((\bar{x}, \bar{y}); \mathrm{gr} F)$. This implies $0_{X^*} \in D^* F(\bar{x}, \bar{y})(y^*)$. Taking into account the relation (7.9) of Lemma 7.9 we obtain (7.11).

(b) Since the proof of the necessary condition is similar to that of the assertion (a)— we only have to use Corollary 7.1(c) instead of Corollary 7.1(b)— we prove here only the sufficient condition. Consider the first case when $\xi_{\bar{y}}$ is convex. By the definition of the coderivative, (7.11) is equivalent to $(0_{X^*}, -y^*) \in N((\bar{x}, \bar{y}); \mathrm{gr} F)$. Since the graph of F is convex, Lemma 7.5(v) implies that the normal cone is equal to the normal cone of convex analysis and we have

$$\langle (0_{X^*}, -y^*), (x - \bar{x}, y - \bar{y}) \rangle \leq 0 \text{ for all } (x, y) \in \mathrm{gr} F,$$

where $\langle \cdot, \cdot \rangle$ denotes the dual pairing. This yields

$$y^*(y - \bar{y}) \geq 0 \text{ for all } y \in F(X). \tag{7.12}$$

Further, since $\xi_{\bar{y}}$ is convex and $y^* \in \partial \xi_{\bar{y}}(\bar{y})$, Lemma 7.5(ii) implies

$$\xi_{\bar{y}}(y) - \xi_{\bar{y}}(\bar{y}) \geq y^*(y - \bar{y}) \text{ for all } y \in F(X).$$

This and (7.12) imply that

$$\xi_{\bar{y}}(y) - \xi_{\bar{y}}(\bar{y}) \geq 0 \text{ for all } y \in F(X).$$

According to Corollary 7.1(d), $(\bar{x}, \bar{y})$ is a globally weakly nondominated solution of (UVP) w.r.t. the ordering map $\mathcal{D}$. Finally, if ℓ is linear and monotone, Lemma 6.7 implies that the functional $\xi_{\bar{y}}$ is convex and we are in the first case. $\square$

Note that according to Remark 6.2(i), the assertion in the above theorem also is necessary for $(\bar{x}, \bar{y})$ to be a locally strongly nondominated solution of the problem (UVP).

Remark 7.11 (Nontriviality of y^* in (7.11)). As $0_{X^*} \in D^* F(\bar{x}, \bar{y})(0_{Y^*})$ holds for any pair $(\bar{x}, \bar{y})$, it is of interest to know whether y^* in (7.11) has to be nonzero. This is the case at least for a locally weakly nondominated solution w.r.t. a variable ordering structure and assuming $\|\ell(y)\|_* > 1$ (hence, int $\mathcal{D}(y) \neq \emptyset$) for all $y \in F(X)$. Then we have in particular $\|\ell(\bar{y})\|_* > 1$, which together with $y^* \in \{\ell(\bar{y})\} + B_{Y^*}$ implies that $\|y^*\|_* > 0$ or $y^* \neq 0_{Y^*}$.

One can derive from Theorem 7.10 various versions of the Fermat rule for the problem (UVP) with the objective map being a single-valued map $f \colon X \to Y$. We illustrate this by formulating just one version for the case when f is strictly differentiable.

Corollary 7.12. *Consider the vector optimization problem*

$$\text{Minimize } f(x) \text{ subject to } x \in X. \tag{7.13}$$

Let $(\bar{x}, \bar{y})$ be a locally nondominated solution w.r.t. the ordering map $\mathcal{D}$ of the problem (7.13) and assume that f is strictly differentiable on X and ℓ is Lipschitz near $\bar{y}$. Then

$$0_{X^*} = [f'(\bar{x})]^*(y^*) \text{ for some } y^* \in \{\ell(\bar{y})\} + B_{Y^*}. \tag{7.14}$$

Proof. According to Lemma 7.6 we have

$$D^* f(\bar{x}, \bar{y})(y^*) = [f'(\bar{x})]^*(y^*).$$

The assertion follows from this and Theorem 7.10. $\square$

Fig. 7.1 Image set $f(X)$ of Example 7.13. In *dark gray* the points $y \in f(X)$ are marked for which there exists some $x \in X$ with (x, y) satisfies the necessary condition (7.14)

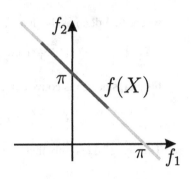

We illustrate Corollary 7.12 by the following example.

Example 7.13. Let the cone-valued map $\mathcal{D}: \mathbb{R}^2 \rightarrow 2^{\mathbb{R}^2}$ be defined by $\mathcal{D}(y) = C(\ell(y))$ with $\ell : \mathbb{R}^2 \rightarrow \mathbb{R}^2$,

$$\ell(y) := \begin{pmatrix} 2 & 1 \\ 1 & 1 \end{pmatrix} y \quad \text{for all } y \in \mathbb{R}^2,$$

$\|\cdot\| = \|\cdot\|_2$ the Euclidean norm, and $f: \mathbb{R}^2 \rightarrow \mathbb{R}^2$,

$$f(x) := \begin{pmatrix} 1 & -1 \\ -1 & 1 \end{pmatrix} x + \begin{pmatrix} 0 \\ \pi \end{pmatrix} \quad \text{for all } x \in \mathbb{R}^2.$$

Then $f(X) = f(\mathbb{R}^2) = \{(t, -t + \pi) \in \mathbb{R}^2 \mid t \in \mathbb{R}\}$, see Fig. 7.1.

The map ℓ is linear and monotone with $\|\ell(y)\|_* > 1$ for all $y \in f(\mathbb{R}^2)$. The map f is strictly differentiable and

$$[f'(x)]^* = \begin{pmatrix} 1 & -1 \\ -1 & 1 \end{pmatrix}^{\mathsf{T}} \quad \text{for all } x \in \mathbb{R}^2.$$

Thus it holds $[f'(x)]^* y^* = (0,0)$ for some $y^* = (y_1^*, y_2^*) \in \mathbb{R}^2$ if and only if $y_1^* = y_2^*$.

For determining those points $(\bar{x}, \bar{y}) \in \mathbb{R}^2 \times f(\mathbb{R}^2)$ which satisfy (7.14) we thus have to search for points $\bar{y} \in f(\mathbb{R}^2)$ and $y^* = (y_1^*, y_2^*) \in \mathbb{R}^2$ with $y_1^* = y_2^*$ and $y^* \in \{\ell(\bar{y})\} + B_{\mathbb{R}^2}$. For any $\bar{y} = (\bar{y}_1, \bar{y}_2) \in f(\mathbb{R}^2)$ it holds $\bar{y}_2 = -\bar{y}_1 + \pi$ and hence we search for $y^*, \bar{y}$ satisfying

$$\begin{pmatrix} y_1^* \\ y_1^* \end{pmatrix} \in \left\{ \begin{pmatrix} \bar{y}_1 + \pi \\ \pi \end{pmatrix} \right\} + B_{\mathbb{R}^2}.$$

Considering the Euclidean norm this is equivalent to searching for $\bar{y}_1 \in \mathbb{R}$ with

$$\min_{y_1^* \in \mathbb{R}} \left\| \begin{pmatrix} \bar{y}_1 + \pi - y_1^* \\ \pi - y_1^* \end{pmatrix} \right\|_2 \leq 1. \tag{7.15}$$

As the minimum in the left hand side in the inequality above is attained in $y_1^* = \pi + \bar{y}_1/2$, (7.15) is equivalent to $|\bar{y}_1| \leq \sqrt{2}$. Hence, a point $(\bar{x}, \bar{y}) \in \mathbb{R}^2 \times f(\mathbb{R}^2)$ satisfies the necessary optimality conditions of Corollary 7.12 if and only if $|\bar{y}_1| \leq \sqrt{2}$. A corresponding y^* is then given by $y^* = (\pi + \bar{y}_1/2, \pi + \bar{y}_1/2)^\top$. The set of points (x, y) satisfying the necessary conditions is thus

$$\left\{ ((x_1, x_2), (x_1 - x_2, \pi - x_1 + x_2)) \in \mathbb{R}^2 \times \mathbb{R}^2 \mid |x_1 - x_2| \leq \sqrt{2} \right\} \qquad (7.16)$$

and all points $(x, y) \in \mathbb{R}^2 \times f(\mathbb{R}^2)$ outside this set cannot be locally nondominated solutions of the problem (7.13) w.r.t. the ordering map $\mathcal{D}$. As all assumptions for the sufficient condition of Theorem 7.10(b) are satisfied in this example the points in the set (7.16) are weakly nondominated solutions of the problem.

For instance, for $(\bar{x}, \bar{y})$ with $\bar{x} := (0, 0)$ and $\bar{y} := f(\bar{x}) = (0, \pi)$ and $y^* := (\pi, \pi)$ we have $y^* \in \{\ell(\bar{y})\} + B_{\mathbb{R}^2} = \{(\pi, \pi)\} + B_{\mathbb{R}^2}$ and $(0, 0) = [f'(\bar{x})]^* y^*$. One can check that $(\bar{x}, \bar{y})$ is in fact a locally nondominated solution of the problem (7.13) w.r.t. the ordering map $\mathcal{D}$.

7.3 Constrained Vector Optimization Problems

In this section we give a Lagrange multiplier rule for the constrained vector optimization problem

$$\text{Minimize } F(x) \text{ subject to } x \in X \text{ and } G(x) \cap C \neq \emptyset, \qquad \text{(CVP)}$$

where F and G are set-valued maps from a Banach space X into Banach spaces Y and Z, respectively, and $C \subset Z$ is a nonempty set (not necessarily a convex cone). Note that one can consider also an additional geometric constraint $x \in \Omega$ but for the sake of simplicity we restrict ourselves here to the case $\Omega = X$. We denote by S the feasible set of (CVP), i.e.

$$S := \{x \in X \mid G(x) \cap C \neq \emptyset\}.$$

The space Y is again assumed to be equipped with a variable ordering structure defined by the ordering map $\mathcal{D}: Y \to 2^Y$ with $\mathcal{D}(y) = \{u \in Y \mid \|u\| \leq \ell(y)(u)\}$ for all $y \in Y$ for some given map $\ell: Y \to Y^*$.

To formulate the Lagrange multiplier rule we need some auxiliary notions and results. A set-valued map F is called *locally pseudo-Lipschitz* around $(\bar{x}, \bar{y}) \in \text{gr } F$ if there exist scalars $r > 0$ and $\gamma \geq 0$ such that

$$(\{\bar{y}\} + r B_Y) \cap F(x) \subset F(x') + \gamma \|x - x'\| B_Y$$

for all $x, x' \in \{\bar{x}\} + rB_X$. For such set-valued maps one has for (x, y) near $(\bar{x}, \bar{y}) \in$ gr F [145],

$$d(y; F(x)) \le (1 + \gamma)d((x; y); \text{gr } F). \qquad (7.17)$$

Further, the set-valued map G is called *metrically regular* around $(\bar{x}, \bar{z}) \in \text{gr } G$ relatively to $X \times C$ if there exist scalars $r > 0$ and $\gamma \ge 0$ such that

$$d((x; z); (X \times C) \cap \text{gr } G) \le \gamma d(z; G(x))$$

for all $(x, z) \in [(\{\bar{x}\} + r \, B_X) \times (\{\bar{z}\} + r \, B_Z)] \cap (X \times C)$. Metric regularity is related to the constraint qualifications for optimization problems with single-valued objective functions [56]. Conditions implying the metric regularity, for instance for Z finite dimensional, are given in [56].

Our argument will make use of the following [56, Proposition 2.6]:

Lemma 7.14. *Suppose that a set-valued map* $G_1 : X \to 2^Y$ *is locally pseudo-Lipschitz around* $(\bar{x}, \bar{y}) \in \text{gr } G_1$ *with some* $\gamma \ge 0$. *Then for any set-valued map* $G_2 : X \to 2^Z$ *with* $\bar{z} \in G_2(\bar{x})$ *and for*

$$\Lambda := \{(x, y, z) \in X \times Y \times Z \mid y \in G_1(x), \; z \in G_2(x)\}$$

one has for (x, y, z) *near* $(\bar{x}, \bar{y}, \bar{z})$

$$d((x, y, z); \Lambda) \le (1 + \gamma)[d((x, y); \text{gr } G_1) + d((x, z); \text{gr } G_2)].$$

We need the following assumption additionally to the standard assumptions as stated above.

Assumption 1. *Let* $\bar{x} \in S$, $\bar{y} \in F(\bar{x})$ *and* $\bar{z} \in G(\bar{x}) \cap C$. *Let the set* C *be closed. Let* ℓ *be Lipschitz near* $\bar{y}$, F *and* G *be closed and locally pseudo-Lipschitz around* $(\bar{x}, \bar{y})$ *and* $(\bar{x}, \bar{z})$ *respectively; and* G *be metrically regular around* $(\bar{x}, \bar{z})$ *relatively to* $X \times C$.

The following theorem gives a Lagrange multiplier rule for the problem (CVP).

Theorem 7.15. *(a) (Necessary condition) Let Assumption 1 be satisfied. If* $(\bar{x}, \bar{y})$ *is a locally nondominated solution of (CVP) w.r.t. the ordering map* $\mathcal{D}$ *then*

$$0_{X^*} \in D^* F(\bar{x}, \bar{y})(y^*) + D^* G(\bar{x}, \bar{z})(z^*) \qquad (7.18)$$

for some $y^* \in \partial \xi_{\bar{y}}(\bar{y}) = \{\ell(\bar{y})\} + B_{Y^*}$ *and* $z^* \in N(\bar{z}; C)$.

(b) (Necessary and sufficient conditions) Suppose that $\|\ell(y)\|_* > 1$ *(hence, int* $\mathcal{D}(y) \ne \emptyset$*) for all* $y \in F(S)$. *If Assumption 1 is satisfied, then (7.18) is necessary for* $(\bar{x}, \bar{y})$ *to be a locally weakly nondominated solution of (CVP) w.r.t. the ordering map* $\mathcal{D}$.

If we assume that the graphs of F and G are convex, $\xi_{\bar{y}}$ is convex (in particular, ℓ is linear and monotone) and C is convex, then (7.18) also is sufficient for $(\bar{x}, \bar{y})$ to be a globally weakly nondominated solution of (UVP) w.r.t. the ordering map $\mathcal{D}$.

Proof. (a) Without loss of generality, we can assume that the same constant $\gamma \geq 0$ is figured in the definitions of the local pseudo-Lipschitz property of F and of G around $(\bar{x}, \bar{y})$ and around $(\bar{x}, \bar{z})$, respectively, and of the metric regularity of G around $(\bar{x}, \bar{z})$ relatively to $X \times C$. Since $(\bar{x}, \bar{y})$ is a locally nondominated solution of (CVP) w.r.t. the ordering map $\mathcal{D}$, $\bar{y}$ is a nondominated element of $F(S \cap U_{\bar{x}})$ w.r.t. $\mathcal{D}$ for some neighborhood $U_{\bar{x}}$ of $\bar{x}$. According to the assumptions there exists a closed ball centered at $\bar{y}$, denoted by $U_{\bar{y}}$ for symmetry with $U_{\bar{x}}$, on which ℓ is Lipschitz. From Lemma 6.6 we deduce that the functional $\xi_{\bar{y}}$ is Lipschitz of rank, say L, on $U_{\bar{y}}$. Corollary 7.1(b) yields that the functional $\xi_{\bar{y}}$ attains its minimum over $F(S \cap U_{\bar{x}})$ at $\bar{y}$. We define a functional $q : X \times Y \times Z \to \mathbb{R}$ by

$$q(x, y, z) := \xi_{\bar{y}}(y) \text{ for all } (x, y, z) \in X \times Y \times Z$$

and

$$\Lambda := \{(x, y, z) \in X \times Y \times Z \mid (x, y) \in (U_{\bar{x}} \times U_{\bar{y}}) \cap \text{gr } F,$$
$$(x, z) \in (X \times C) \cap \text{gr } G\}.$$

One can easily check that $\Lambda \subset U_{\bar{x}} \times U_{\bar{y}} \times Z$ and $(\bar{x}, \bar{y}, \bar{z})$ is a minimizer of the functional q on Λ. Note that q is Lipschitz of rank L on $U_{\bar{x}} \times U_{\bar{y}} \times Z$. Then by the Clarke penalization, see Lemma 7.7, $(\bar{x}, \bar{y}, \bar{z})$ is a minimizer (and hence, a local minimizer) of the functional $q(\cdot) + L\, d(\cdot; \Lambda)$ on $U_{\bar{x}} \times U_{\bar{y}} \times Z$. Therefore, we can assume that

$$0 = \xi_{\bar{y}}(\bar{y}) + L\, d((\bar{x}, \bar{y}, \bar{z}); \Lambda) \leq \xi_{\bar{y}}(y) + L\, d((x, y, z); \Lambda) \qquad (7.19)$$

for (x, y, z) near $(\bar{x}, \bar{y}, \bar{z})$. We apply Lemma 7.14 to the maps $G_1 : X \to 2^Y$ defined by

$$G_1(x) := \begin{cases} F(x) \cap U_{\bar{y}} & \text{if } x \in U_{\bar{x}}, \\ \emptyset & \text{else,} \end{cases}$$

and $G_2 : X \to 2^Z$,

$$G_2(x) := G(x) \cap C.$$

G_1 is locally pseudo-Lipschitz around $(\bar{x}, \bar{y})$ and the set Λ as defined above equals

$$\Lambda = \{(x, y, z) \in X \times Y \times Z \mid y \in G_1(x),\ z \in G_2(x)\}.$$

Thus for (x, y, z) near $(\bar{x}, \bar{y}, \bar{z})$ we have

$$d((x, y, z); \Lambda) \leq (1 + \gamma)[d((x, y); (U_{\bar{x}} \times U_{\bar{y}}) \cap \operatorname{gr} F) \tag{7.20}$$
$$+d((x, z); (X \times C) \cap \operatorname{gr} G)].$$

Further, the metric regularity of G around $(\bar{x}, \bar{z})$ relatively to $X \times C$ yields that

$$d((x, z); (X \times C) \cap \operatorname{gr} G) \leq \gamma d(z; G(x)) \tag{7.21}$$

for $(x, z) \in X \times C$ near $(\bar{x}, \bar{z})$ and the local pseudo-Lipschitz property of G around $(\bar{x}, \bar{z})$ (see (7.17)) yields

$$d(z; G(x)) \leq (1 + \gamma)d((x, z); \operatorname{gr} G) \tag{7.22}$$

for (x, z) near $(\bar{x}, \bar{z})$. Combining (7.19)–(7.22) gives

$$0 \leq \xi_{\bar{y}}(y) + L\, d((x, y, z); \Lambda)$$
$$\leq \xi_{\bar{y}}(y) + L\,(1 + \gamma)[d((x, y); (U_{\bar{x}} \times U_{\bar{y}}) \cap \operatorname{gr} F)$$
$$+d((x, z); (X \times C) \cap \operatorname{gr} G)]$$
$$\leq \xi_{\bar{y}}(y) + L(1 + \gamma)[d((x, y); (U_{\bar{x}} \times U_{\bar{y}}) \cap \operatorname{gr} F)$$
$$+\gamma(1 + \gamma)d((x, z); \operatorname{gr} G)]$$

for $(x, y, z) \in X \times Y \times C$ near $(\bar{x}, \bar{y}, \bar{z})$. This means that $(\bar{x}, \bar{y}, \bar{z})$ is a local minimizer of the functional $\tilde{q}$ over $X \times Y \times C$ that is defined by $\tilde{q}: X \times Y \times Z \to \mathbb{R}$,

$$\tilde{q}(x, y, z) := \xi_{\bar{y}}(y) + (1 + \gamma)\, L\, d((x, y); (U_{\bar{x}} \times U_{\bar{y}}) \cap \operatorname{gr} F)$$
$$+\gamma\,(1 + \gamma)^2\, L\, d((x, z); \operatorname{gr} G).$$

It is clear that $\tilde{q}$ is Lipschitz near $(\bar{x}, \bar{y}, \bar{z})$. Applying Corollary 7.8, we obtain that

$$0_{X^* \times Y^* \times Z^*} \in \partial\tilde{q}(\bar{x}, \bar{y}, \bar{z}) + N((\bar{x}, \bar{y}, \bar{z}); X \times Y \times C)$$
$$= \partial\tilde{q}(\bar{x}, \bar{y}, \bar{z}) + \{0_{X^*}\} \times \{0_{Y^*}\} \times N(\bar{z}; C).$$

By Lemma 7.5(iv)–(v), there exist

$$y_1^* \in \partial\xi_{\bar{y}}(\bar{y}),$$

$$(x_2^*, y_2^*) \in (1 + \gamma)L\, \partial d((\bar{x}, \bar{y}); (U_{\bar{x}} \times U_{\bar{y}}) \cap \operatorname{gr} F)$$
$$\subset N((\bar{x}, \bar{y}); (U_{\bar{x}} \times U_{\bar{y}}) \cap \operatorname{gr} F)$$
$$= N((\bar{x}, \bar{y}); \operatorname{gr} F),$$

$$(x_3^*, z_3^*) \in \gamma(1 + \gamma)^2 L\, \partial d((\bar{x}, \bar{z}); \operatorname{gr} G) \subset N((\bar{x}, \bar{z}); \operatorname{gr} G),$$

$$z_4^* \in N(\bar{z}; C),$$

such that

$$0_{X^*} = x_2^* + x_3^*, \qquad 0_{Y^*} = y_1^* + y_2^* \quad \text{and} \quad 0_{Z^*} = z_3^* + z_4^*.$$

Taking into account the relation (7.9) of Lemma 7.9, we deduce that

$$y_1^* \in \partial \xi_{\bar{y}}(\bar{y}) = \{\ell(\bar{y})\} + B_{Y^*}$$

and putting $y^* := y_1^* = -y_2^*$ and $z^* := z_4^* = -z_3^*$, we obtain $x_2^* \in D^* F(\bar{x}, \bar{y})(y^*)$ and $x_3^* \in D^* G(\bar{x}, \bar{z})(z^*)$ and thus

$$0_{X^*} \in D^* F(\bar{x}, \bar{y})(y^*) + D^* G(\bar{x}, \bar{z})(z^*)$$

and (7.18) holds.

(b) Since the proof of the necessary condition is similar to that of the assertion (a), we prove only the sufficient condition. Suppose that the graphs of F, G and the set C are convex and (7.18) holds. Consider the first case when $\xi_{\bar{y}}$ is convex. By (7.18) one can find elements $x_1^*, x_2^* \in X^*$, $y^* \in \partial \xi_{\bar{y}}(\bar{y})$ and $z^* \in N(\bar{z}; C)$ such that $(x_1^*, -y^*) \in N((\bar{x}, \bar{y}); \mathrm{gr}\, F)$, $(x_2^*, -z^*) \in N((\bar{x}, \bar{z}); \mathrm{gr}\, G)$ and

$$x_1^* + x_2^* = 0_{X^*}. \tag{7.23}$$

Since the graphs of F, G and the set C are convex, according to the definition of the normal cone of convex analysis, see Lemma 7.5(v), we have

$$\langle (x_1^*, -y^*), (x, y) - (\bar{x}, \bar{y}) \rangle \leq 0 \text{ for all } (x, y) \in \mathrm{gr}\, F \tag{7.24}$$

$$\langle (x_2^*, -z^*), (x, z) - (\bar{x}, \bar{z}) \rangle \leq 0 \text{ for all } (x, z) \in \mathrm{gr}\, G \tag{7.25}$$

and

$$\langle z^*, z - \bar{z} \rangle \leq 0 \text{ for all } z \in C. \tag{7.26}$$

Summarizing (7.24)–(7.26) and taking account of (7.23) we obtain

$$\langle y^*, y - \bar{y} \rangle \geq 0 \text{ for all } y \in F(S). \tag{7.27}$$

Further, since $\xi_{\bar{y}}$ is convex and $y^* \in \partial \xi_{\bar{y}}(\bar{y})$, the definition of the subdifferential in the sense of convex analysis (Lemma 7.5(ii)) yields

$$\langle y^*, y - \bar{y} \rangle \leq \xi_{\bar{y}}(y) - \xi_{\bar{y}}(\bar{y}),$$

which together with (7.27) implies

$$\xi_{\bar{y}}(y) - \xi_{\bar{y}}(\bar{y}) \geq 0 \text{ for all } y \in F(S).$$

Corollary 7.1(d) then implies that $(\bar{x}, \bar{y})$ is a globally weakly nondominated solution of (CVP) w.r.t. the ordering map $\mathcal{D}$. Finally, if ℓ is linear and monotone, Lemma 6.7 implies that the functional $\xi_{\bar{y}}$ is convex and we are in the first case.

$\square$

Similar as in the unconstrained case, see Remark 7.11, the element y^* in (7.18) has to be nonzero under the assumption that $\|\ell(y)\|_* > 1$ (hence, $\operatorname{int}(\mathcal{D}(y)) \neq \emptyset$) for all $y \in F(S)$.

When both the objective and the constraint maps are single-valued, one can apply techniques from scalar optimization to deduce a Lagrange multiplier rule for (CVP) w.r.t. a variable ordering structure. We illustrate this by considering the following Lipschitz finite-dimensional case. In the theorems and the example below, the subdifferential is understood in the sense of Clarke. We use in the proof the following result by Clarke.

Theorem 7.16 ([34, Theorem 6.1.1]). *Let* $h: \mathbb{R}^n \to \mathbb{R}$ *and also* $g = (g_1, \ldots, g_s): \mathbb{R}^n \to \mathbb{R}^s$ *be single-valued maps. Let* $\bar{x}$ *be a minimal solution of the scalar-valued optimization problem*

Minimize $h(x)$ *subject to* $g_j(x) \leq 0$, $j = 1, \ldots, s$ *and* $x \in \mathbb{R}^n$.

Then there exist scalars $\mu \geq 0$, $\lambda_j \geq 0$ ($j = 1, \ldots, s$) *not all zero such that*

$$\sum_{j=1}^{s} \lambda_j g_j(\bar{x}) = 0$$

and

$$0_{\mathbb{R}^n} \in \mu \, \partial h(\bar{x}) + \sum_{j=1}^{s} \lambda_j \partial g_j(\bar{x}).$$

Theorem 7.17. *Let* $f: \mathbb{R}^n \to \mathbb{R}^m$ *and* $g = (g_1, \ldots, g_s): \mathbb{R}^n \to \mathbb{R}^s$ *be single-valued maps. Consider the vector optimization problem*

Minimize $f(x)$ *subject to* $g_j(x) \leq 0$, $j = 1, \ldots, s$ *and* $x \in \mathbb{R}^n$.   (7.28)

Let $(\bar{x}, \bar{y})$ *(here,* $\bar{y} = f(\bar{x})$*) be a locally nondominated solution w.r.t. the ordering map* $\mathcal{D}$ *of the problem (7.28). Assume that* f *and* g_j ($j = 1, \ldots, s$) *are Lipschitz near* $\bar{x}$ *and* ℓ *is Lipschitz near* $\bar{y}$. *Then there exist scalars* $\mu \geq 0$, $\lambda_j \geq 0$ ($j = 1, \ldots, s$) *not all zero such that*

$$\sum_{j=1}^{s} \lambda_j g_j(\bar{x}) = 0$$ (7.29)

and

$$0_{\mathbb{R}^n} \in \mu \, \partial(\xi_{\bar{y}} \circ f)(\bar{x}) + \sum_{j=1}^{s} \lambda_j \partial g_j(\bar{x}). \tag{7.30}$$

Proof. Since $(\bar{x}, \bar{y})$ is a locally nondominated solution of (7.28) w.r.t. the ordering map $\mathcal{D}$, $\bar{y}$ is a nondominated element of $f(S \cap U_{\bar{x}})$ w.r.t. $\mathcal{D}$ for some neighborhood $U_{\bar{x}}$ of $\bar{x}$, where

$$S := \{x \in \mathbb{R}^n \mid g_j(x) \leq 0, \ j = 1, \ldots, s\}.$$

Corollary 7.1 yields that $\xi_{\bar{y}}$ attains its minimum over $f(S \cap U_{\bar{x}})$ at $\bar{y}$. Therefore, $\bar{x}$ is a local minimizer of the scalar-valued optimization problem

Minimize $(\xi_{\bar{y}} \circ f)(x)$ subject to $g_j(x) \leq 0, \ j = 1, \ldots, s$ and $x \in \mathbb{R}^n$. (7.31)

One can easily show that the composition map $\xi_{\bar{y}} \circ f : \mathbb{R}^n \to \mathbb{R}$ is Lipschitz near $\bar{x}$ using Lemma 6.6. The assertion now follows from Theorem 7.16 applied to the problem (7.31). □

We end this section with an example illustrating Theorem 7.17.

Example 7.18. Let $\mathbb{R}^2$ be the Euclidean space and the cone-valued map $\mathcal{D} : \mathbb{R}^2 \to 2^{\mathbb{R}^2}$ be defined by $\mathcal{D}(y) = C(\ell(y))$ with $\ell : \mathbb{R}^2 \to \mathbb{R}^2$,

$$\ell(y) := \frac{1}{\pi} \begin{pmatrix} 2 & 1 \\ -1 & -1 \end{pmatrix} y \ \text{ for all } \ y \in \mathbb{R}^2.$$

Consider the problem (7.28) with the maps $f : \mathbb{R}^2 \to \mathbb{R}^2$, $g : \mathbb{R}^2 \to \mathbb{R}^2$ defined by

$$f(x_1, x_2) := \left(x_1^2, x_2^2\right)^{\top} \ \text{ for all } \ (x_1, x_2) \in \mathbb{R}^2,$$

$$g(x_1, x_2) := \begin{pmatrix} \pi - (x_1^2 + x_2^2) \\ x_1^2 + x_2^2 - 2\pi \end{pmatrix} \ \text{ for all } \ (x_1, x_2) \in \mathbb{R}^2.$$

Observe that f and g are strictly differentiable (hence, they are locally Lipschitz by Lemma 7.4) and that ℓ is linear and Lipschitz on $\mathbb{R}^2$. We start by calculating the subdifferentials figured in (7.30). For some $x, \bar{x} \in \mathbb{R}^2$ and $\bar{y} := f(\bar{x})$ we obtain

$$(\eta_{\bar{y}} \circ f)(x) = \ell(f(x))(f(x) - f(\bar{x}))$$

$$= \frac{1}{\pi} \begin{pmatrix} 2x_1^2 + x_2^2 \\ -x_1^2 - x_2^2 \end{pmatrix}^{\top} \begin{pmatrix} x_1^2 - \bar{x}_1^2 \\ x_2^2 - \bar{x}_2^2 \end{pmatrix}$$

$$= \frac{1}{\pi} \left(2x_1^4 - 2x_1^2 \bar{x}_1^2 - \bar{x}_1^2 x_2^2 + x_1^2 \bar{x}_2^2 - x_2^4 + x_2^2 \bar{x}_2^2\right)$$

Fig. 7.2 Image set $f(S)$ of Example 7.18

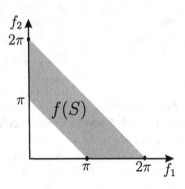

and thus, based on Lemma 7.5(i),

$$\partial(\eta_{\bar{y}} \circ f)(x) = \left\{ \frac{1}{\pi} \begin{pmatrix} 8x_1^3 - 4x_1\bar{x}_1^2 + 2x_1\bar{x}_2^2 \\ -2\bar{x}_1^2 x_2 - 4x_2^3 + 2x_2\bar{x}_2^2 \end{pmatrix} \right\}.$$

Lemma 7.5(iv) implies, since $\xi_{\bar{y}}(y) = \eta_{\bar{y}}(y) + \|y - \bar{y}\|$,

$$\partial(\xi_{\bar{y}} \circ f)(\bar{x}) = \partial(\eta_{\bar{y}} \circ f)(\bar{x}) + \partial\|f(\cdot) - f(\bar{x})\| \bigg|_{x=\bar{x}}.$$

With Lemma 7.5(vi), the chain rule given in Theorem 2.3.10 and Remark 2.3.11 in [34] on the meaning of the chain rule, we obtain

$$\partial(\xi_{\bar{y}} \circ f)(\bar{x}) = \left\{ \frac{1}{\pi} \begin{pmatrix} 4\bar{x}_1^3 + 2\bar{x}_1\bar{x}_2^2 \\ -2\bar{x}_2^3 - 2\bar{x}_1^2\bar{x}_2 \end{pmatrix} \right\} + [f'(\bar{x})]^* B_{\mathbb{R}^2}. \qquad (7.32)$$

Here,

$$[f'(\bar{x})]^* B_{\mathbb{R}^2} = \left\{ y \in \mathbb{R}^2 \,\bigg|\, y = \begin{pmatrix} 2\bar{x}_1 & 0 \\ 0 & 2\bar{x}_2 \end{pmatrix} x, \; x \in B_{\mathbb{R}^2} \right\}.$$

Further,

$$\partial g_1(\bar{x}) = \{(-2\bar{x}_1, -2\bar{x}_2)^\top\} \quad \text{and} \quad \partial g_2(\bar{x}) = \{(2\bar{x}_1, 2\bar{x}_2)^\top\}.$$

The set $f(S)$ with $S := \{(x_1, x_2) \in \mathbb{R}^2 \mid g_i(x_1, x_2) \leq 0, \; i = 1, 2\}$ is given by

$$f(S) = \{(y_1, y_2) \in \mathbb{R}_+^2 \mid \pi \leq y_1 + y_2 \leq 2\pi\},$$

see Fig. 7.2.

In the following we choose three points $(\bar{x}, \bar{y}) \in S \times f(S)$ and check whether they satisfy the necessary optimality conditions presented in Theorem 7.17, i.e., whether they are candidates for being locally nondominated solutions of (7.28) w.r.t. $\mathcal{D}$.

(i) Let $\bar{x} := (\sqrt{2\pi}, 0)$ then $\bar{y} = (2\pi, 0)$ and $g_1(\bar{x}) = -\pi$, $g_2(\bar{x}) = 0$ and thus condition (7.29) implies $\lambda_1 = 0$. Using (7.32) we need to check if there are $\mu \geq 0$, $\lambda_2 \geq 0$ satisfying

$$0_{\mathbb{R}^2} \in \mu \left(\left\{ \begin{pmatrix} 8\sqrt{2\pi} \\ 0 \end{pmatrix} \right\} + \begin{pmatrix} 2\sqrt{2\pi} & 0 \\ 0 & 0 \end{pmatrix} B_{\mathbb{R}^2} \right) + \lambda_2 \left\{ \begin{pmatrix} 2\sqrt{2\pi} \\ 0 \end{pmatrix} \right\}.$$

This implies that there is some $(b_1, b_2) \in B_{\mathbb{R}^2}$ such that

$$2\sqrt{2\pi}(\mu(4 + b_1) + \lambda_2) = 0$$

and this only holds for $\mu = \lambda_2 = 0$. Thus $(\bar{x}, \bar{y})$ is not a locally nondominated solution of (7.28) w.r.t. $\mathcal{D}$.

(ii) Let $\bar{x} := (\sqrt{\pi}, 0)$ then $\bar{y} = (\pi, 0)$ and $g_1(\bar{x}) = 0$, $g_2(\bar{x}) = -\pi$ and thus $\lambda_2 = 0$. As $0_{\mathbb{R}^2} \in \partial \| f(\cdot) - f(\bar{x}) \| \Big|_{x=\bar{x}}$, we have

$$(4\sqrt{\pi}, 0)^{\top} \in \partial(\xi_{\bar{y}} \circ f)(\bar{x}).$$

With $\partial g_1(\bar{x}) = \{(-2\sqrt{\pi}, 0)^{\top}\}$, we obtain for $\mu = 1$ and $\lambda_1 = 2$ that (7.30) holds. Hence, $(\bar{x}, \bar{y})$ satisfies the necessary optimality conditions of Theorem 7.17. Nevertheless it is not a locally nondominated solution, because $\bar{y}$ is not a nondominated element of $f(S \cap U_{\bar{x}})$ for any neighborhood $U_{\bar{x}}$ of $\bar{x}$: for any $\varepsilon \in [0, (3 - \sqrt{2})\pi]$ we define $x^{\varepsilon} := (\sqrt{\pi - \varepsilon}, \sqrt{\varepsilon}) \in S$ and $y^{\varepsilon} := f(\sqrt{\pi - \varepsilon}, \sqrt{\varepsilon}) = (\pi - \varepsilon, \varepsilon)$. Then $\lim_{\varepsilon \to 0} x^{\varepsilon} = \bar{x}$ and

$$\bar{y} - y^{\varepsilon} = (\varepsilon, -\varepsilon)^{\top} \in \mathcal{D}(y^{\varepsilon}) = C(\ell(y^{\varepsilon}))$$

$$= \left\{ u \in \mathbb{R}^2 \mid \|u\| \leq \tfrac{1}{\pi}(2\pi - \varepsilon, -\pi)u \right\}.$$

(iii) Let $\bar{x} := (0, \sqrt{2\pi})$ then $\bar{y} = (0, 2\pi)$ and $g_1(\bar{x}) = -\pi$, $g_2(\bar{x}) = 0$, and thus $\lambda_1 = 0$. As $0_{\mathbb{R}^2} \in \partial \| f(\cdot) - f(\bar{x}) \| \Big|_{x=\bar{x}}$, we have

$$(0, -4\sqrt{2\pi})^{\top} \in \partial(\xi_{\bar{y}} \circ f)(\bar{x}).$$

Further, $\partial g_2(\bar{x}) = \{(0, 2\sqrt{2\pi})^{\top}\}$. Then, for $\mu = 1$ and $\lambda_2 = 2$ the necessary optimality condition (7.30) is satisfied. We claim that $(\bar{x}, \bar{y})$ is in fact a globally nondominated solution of (7.28) w.r.t. $\mathcal{D}$. For showing this, we assume $\bar{y}$ is not a nondominated element of $f(S)$ w.r.t. $\mathcal{D}$. Then there is some $y \in f(S)$, $y \neq \bar{y}$, with $\bar{y} - y \in \mathcal{D}(y) = C(\ell(y))$, i.e. with

$$\|\bar{y} - y\| \le l(y)^{\top}(\bar{y} - y) = \frac{1}{\pi}\begin{pmatrix} 2y_1 + y_2 \\ -y_1 - y_2 \end{pmatrix}^{\top}\begin{pmatrix} -y_1 \\ 2\pi - y_2 \end{pmatrix}$$

$$= \frac{1}{\pi}(\underbrace{y_2}_{\ge 0} \underbrace{(y_2 - 2\pi)}_{\le 0} - \underbrace{y_1(2y_1 + 2\pi)}_{\ge 0}) \le 0.$$

This is a contradiction, as $\bar{y} \ne y$.

7.4   Notes on the Literature

The Ioffe approximate normal cone and the approximate subdifferential were defined by Ioffe in [88] for a lower semicontinuous map under the name G-nucleus of the G-normal cone and G-nucleus of the G-subdifferential. For the approximate coderivative see Ioffe and Penot [89]. We refer to the book by Clarke [34] for the Clarke generalized subdifferential and normal cone. The theory of the Mordukhovich subdifferential and coderivative was provided by Mordukhovich in [117–120]. He introduced the notion of a coderivative for set-valued maps based on a non-specified normal cone in [117] which can thus also be applied to the normal cone in the sense of Clarke which is then denoted the Clarke's coderivative [118]. The Mordukhovich coderivative and the Mordukhovich subdifferential are also called limiting Fréchet coderivative and limiting Fréchet normal cone.

Characterizations of strict differentiability (Lemma 7.4) and the results regarding the Clarke penalization (Lemma 7.7 and Corollary 7.8) can be found in the book by Clarke [34], see [34, Propositions 2.2.1, 2.4.3 and p. 52]. The collection in the Lemmas 7.5 and 7.6 is taken from [34, 56, 88, 117–120] by Clarke, El Abdouni, Thibault, Ioffe and Mordukhovich. The formulation of the subdifferential in Lemma 7.5(vi) is due to Mordukhovich and Lewis (private communication). The subdifferentials and the strict derivative of the scalarization functionals in Lemma 7.9 are due to Eichfelder and Ha [52].

For an overview over the first necessary optimality conditions for constrained set-optimization problems of the type (CVP) in partially ordered spaces see [56]. We refer to Aubin and Ekeland [7], see also El Abdouni and Thibault [56, 145], for the notion of pseudo-Lipschitz, and for (7.17). For the notion of metrically regularity we refer to El Abdouni and Thibault [56, Definition 2.1], [145]. Lemma 7.14 in the form presented here equals Proposition 2.6 in [56] by El Abdouni and Thibault. In a similar form it was already stated by Thibault in [145, Proposition 1.2] and obtained for Lipschitz set-valued maps by Clarke [33]. The proof of Theorem 7.15 is similar to that of Theorem 3.7 in [56]. The optimality conditions of Fermat and Lagrange type for variable ordering structures as presented in Sects. 7.2 and 7.3 were first given by Eichfelder and Ha in [52]. Bao and Mordukhovich also developed necessary conditions for nondominated elements for variable ordering structures but without the assumption that the images of the ordering map are Bishop-Phelps cones in [11].

Chapter 8
Duality Results

In vector optimization as well as in scalar optimization it is well known that, under appropriate assumptions, a maximization problem can be associated with a minimization problem [94]. In this chapter we associate such a dual problem with a vector optimization problem with a variable ordering structure. This relates (weakly) nondominated elements and solutions of the original problem with (weakly) maximal elements of the dual problem, or, in Sect. 8.2, weakly minimal elements and solutions and weakly max-nondominated elements.

We start by presenting duality concepts in a very general way for sets, and in the subsequent sections we collect duality results for vector optimization problems based on linear and nonlinear scalarizations. For the notions of maximal and max-nondominated elements of a set w.r.t. $\mathcal{D}$ see page 39.

In this chapter, if not noted otherwise, Y is a real linear space equipped with a variable ordering structure defined by an ordering map $\mathcal{D}: Y \rightarrow 2^Y$ with $\mathcal{D}(y)$ a pointed convex cone for all $y \in Y$ and A is a nonempty subset of Y.

8.1   General Duality

In a partially ordered space Y with an ordering cone K, one can define a *weak duality* condition [64] for a dual set $H \subset Y$ by

$$A \cap (H - (K \setminus \{0_Y\})) = \emptyset.$$

This condition, which is equivalent to $(A + K \setminus \{0_Y\}) \cap H = \emptyset$, is also known as *dominance requirement*, as it ensures that the elements of the dual set are not dominated by the elements of the primal set. For the case of a variable ordering structure, this condition can be formulated in two ways:

$$A \cap (\{y\} - (\mathcal{D}(y) \setminus \{0_Y\})) = \emptyset \ \text{ for all } y \in H \tag{8.1}$$

G. Eichfelder, *Variable Ordering Structures in Vector Optimization*, Vector Optimization, DOI 10.1007/978-3-642-54283-1_8, © Springer-Verlag Berlin Heidelberg 2014

or

$$(\{y\} + (\mathcal{D}(y) \setminus \{0_Y\})) \cap H = \emptyset \text{ for all } y \in A. \tag{8.2}$$

If, additionally, $0_Y \in \mathrm{cl}(A - H)$, i.e. A and H touch each other, then it is said that *strong duality* holds [71, p. 155]. If $A \cap H \neq \emptyset$ then this additional assumption is satisfied.

In the following we introduce two different dual sets, Q and T. The first satisfies the weak duality (8.2) and we obtain duality and converse duality results relating the optimal elements of the original and the dual set. The second dual set also satisfies (8.2) and under weak assumptions also (8.1), but the optimal elements are only related under additional assumptions.

To the so-called primal set $A \subset Y$, we first define a dual set based on the set

$$M := \bigcup_{y \in A} \{y\} + \mathcal{D}(y),$$

which we have already examined in Lemma 2.38. We define the dual set by

$$Q := Y \setminus \tilde{M} \quad \text{with} \quad \tilde{M} := \bigcup_{y \in A} \{y\} + (\mathcal{D}(y) \setminus \{0_Y\}).$$

Example 8.1. Let Y, A and $\mathcal{D}$ be defined as in Example 2.10, i.e. let Y be the Euclidean space $\mathbb{R}^2$, $A := \{y \in \mathbb{R}^2 \mid \|y\| \leq 1\}$ and $\mathcal{D}: \mathbb{R}^2 \to 2^{\mathbb{R}^2}$,

$$\mathcal{D}(y) := \begin{cases} \mathbb{R}^2_+ & \text{for } y \in \mathbb{R}^2 \setminus \{(0, -1)\}, \\ \{(z_1, z_2) \in \mathbb{R}^2 \mid z_1 + z_2 \geq 0, z_1 \geq 0\} & \text{for } y = (0, -1). \end{cases}$$

For the set M see Example 2.39 and Fig. 2.7. The set Q is given by

$$\begin{aligned} Q = &\{(y_1, y_2) \in \mathbb{R}^2 \mid y_1 < -1\} \\ &\cup \{(y_1, y_2) \in \mathbb{R}^2 \mid y_1 \in [-1, 0], \ y_2 \leq -\sqrt{1 - y_1^2}\} \\ &\cup \{(y_1, y_2) \in \mathbb{R}^2 \mid y_1 > 0, \ y_2 < -1 - y_1\}; \end{aligned}$$

see Fig. 8.1.

Lemma 8.2. *For the dual set $H := Q$ the weak duality (8.2) is satisfied.*

Proof. For $y \in A$ we have $\{y\} + (\mathcal{D}(y) \setminus \{0_Y\}) \subset \tilde{M}$. As $Q = Y \setminus \tilde{M}$, we arrive at $(\{y\} + (\mathcal{D}(y) \setminus \{0_Y\}) \cap Q = \emptyset$ for all $y \in A$. $\square$

Lemma 8.3. *If $\bar{y} \in A \cap Q$, then $\bar{y}$ is a nondominated element of A w.r.t. $\mathcal{D}$ and $\bar{y}$ is a maximal element of Q w.r.t. $\mathcal{D}$.*

Proof. As $\bar{y} \in Q = Y \setminus \tilde{M}$, and hence $\bar{y} \notin \tilde{M}$, we immediately obtain that $\bar{y}$ is nondominated w.r.t. $\mathcal{D}$ for the set A. Since $Q = Y \setminus \tilde{M}$ it holds that $Q \cap \tilde{M} = \emptyset$

Fig. 8.1 Set Q
of Example 8.1, cf. [46]

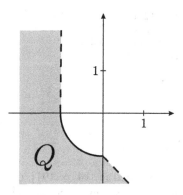

and hence $Q \cap (\{\bar{y}\} + (\mathcal{D}(\bar{y}) \setminus \{0_Y\})) = \emptyset$. Thus $\bar{y}$ is a maximal element of Q w.r.t. $\mathcal{D}$. $\qquad\square$

Now we can formulate the following duality theorem:

Theorem 8.4. *If $\bar{y} \in A$ is a nondominated element of A w.r.t. $\mathcal{D}$, then $\bar{y}$ is a maximal element of Q w.r.t. $\mathcal{D}$.*

Proof. Let $\bar{y} \in A$ be a nondominated element of A w.r.t. $\mathcal{D}$ and assume, that $\bar{y} \notin Q$. Then $\bar{y} \in \tilde{M}$, which is a contradiction to $\bar{y}$ being a nondominated element of A w.r.t. $\mathcal{D}$. Thus $\bar{y} \in A \cap Q$ and the assertion follows with Lemma 8.3. $\qquad\square$

Together with Lemma 2.38(i) we conclude:

Corollary 8.5. *If $\bar{y} \in M$ is a nondominated element of M w.r.t. $\mathcal{D}$, then $\bar{y}$ is a maximal element of Q w.r.t. $\mathcal{D}$.*

The next theorem is the associated converse duality theorem.

Theorem 8.6. *If $Y \setminus M$ is algebraically open, then every maximal element of the set Q w.r.t. $\mathcal{D}$ is also a nondominated element of the set A w.r.t. $\mathcal{D}$.*

Proof. Let $\bar{y}$ be an arbitrary maximal element of Q w.r.t. $\mathcal{D}$, i.e. $(\{\bar{y}\} + \mathcal{D}(\bar{y})) \cap Q = \{\bar{y}\}$. We assume that $\bar{y} \notin M$. As $Y \setminus M$ is algebraically open, for every $d \in \mathcal{D}(\bar{y}) \setminus \{0_Y\}$ a scalar $\lambda > 0$ exists with

$$\bar{y} + \lambda d \in Y \setminus M \subset Q.$$

As

$$\bar{y} + \lambda d \in \{\bar{y}\} + (\mathcal{D}(\bar{y}) \setminus \{0_Y\}),$$

this is a contradiction to the maximality of $\bar{y}$. Thus $\bar{y} \in M$ and due to $\bar{y} \notin \tilde{M}$, we obtain $\bar{y} \in A$. Together with Lemma 8.3, we conclude that $\bar{y}$ is a nondominated element of A w.r.t. $\mathcal{D}$. $\qquad\square$

Fig. 8.2 (**a**) Sets A and Q
of Example 8.8. (**b**) Sets
A and Q of Example 8.9,
cf. [46]

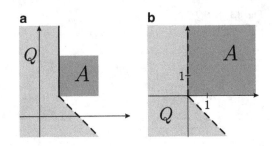

As stated in [94, p. 191], the set $Y \setminus M$ is, for instance, algebraically open if the set M is convex and algebraically closed. But in general, we need no convexity assumptions in Theorem 8.6.

Example 8.7. We consider the set A and the ordering map $\mathcal{D}$ as defined in Example 2.10; compare Example 8.1. The set of all maximal elements of Q w.r.t. $\mathcal{D}$ is

$$\left\{ (y_1, y_2) \in \mathbb{R}^2 \mid y_1 \in [-1, 0], \ y_2 = -\sqrt{1 - y_1^2} \right\}.$$

This set equals the set of nondominated elements of A w.r.t. $\mathcal{D}$. The set $Y \setminus M$ is algebraically open.

Example 8.8. We consider the set $A = [1, 3] \times [1, 3]$ and $\mathcal{D}: \mathbb{R}^2 \to 2^{\mathbb{R}^2}$ given by

$$\mathcal{D}(y) := \begin{cases} \mathbb{R}_+^2, & \text{if } y_1 > 1, \\ \{(z_1, z_2) \in \mathbb{R}^2 \mid z_1 + z_2 \geq 0, \ z_1 - z_2 \geq 0\}, & \text{else.} \end{cases}$$

The sets $\tilde{M}$, M and Q in this example are

$$\tilde{M} = \{(y_1, y_2) \in \mathbb{R}^2 \mid y_1 > 1, \ y_2 \geq 2 - y_1\},$$
$$M = \tilde{M} \cup \{(y_1, y_2) \in \mathbb{R}^2 \mid y_1 = 1, \ y_2 \in [1, 3]\},$$
$$Q = \{(y_1, y_2) \in \mathbb{R}^2 \mid y_1 \leq 1 \ \vee \ (y_1 > 1 \ \wedge \ y_1 + y_2 < 2)\};$$

see Fig. 8.2a.

The set of nondominated elements of A w.r.t. the ordering map $\mathcal{D}$ is $\{(y_1, y_2) \in \mathbb{R}^2 \mid y_1 = 1, \ y_2 \in [1, 3]\}$, and the set of maximal elements of Q w.r.t. $\mathcal{D}$ is $\{(y_1, y_2) \in \mathbb{R}^2 \mid y_1 = 1, \ y_2 \geq 1\}$. The set $Q \cap A$ equals the set of nondominated elements of A w.r.t. $\mathcal{D}$, which is a strict subset of the set of maximal elements of Q w.r.t. $\mathcal{D}$. Thus not all maximal elements of Q w.r.t. $\mathcal{D}$ refer to a nondominated element of A w.r.t. $\mathcal{D}$. The set $Y \setminus M$ is not algebraically open.

Example 8.9. Let $A = \mathbb{R}_+^2$ and $\mathcal{D}: \mathbb{R}^2 \to 2^{\mathbb{R}^2}$ be given by

$$\mathcal{D}(y) := \begin{cases} \{(z_1, z_2) \in \mathbb{R}^2 \mid z_1 \geq 0, \ z_1 + z_2 \geq 0\}, & \text{if } y_2 \geq 0, \\ \mathbb{R}_+^2, & \text{else.} \end{cases}$$

Then $M = \{(y_1, y_2) \in \mathbb{R}^2 \mid y_1 \geq 0, \ y_1 + y_2 \geq 0\}$ and $\tilde{M} = M \setminus \{(0,0)\}$. The set $Y \setminus M$ is algebraically open and $Q = (\mathbb{R}^2 \setminus M) \cup \{(0,0)\}$; see Fig. 8.2b. The set Q has the unique maximal element $(0,0)$ w.r.t. $\mathcal{D}$, which is also the unique nondominated element of A w.r.t. $\mathcal{D}$. Note that, for instance, $(1,-1) \notin Q$.

Also another dual set can be defined which satisfies under weak assumptions both dominance requirements (8.1) and (8.2) but for which duality results as above can only be obtained under additional assumptions. We present this dual set for completeness. Define the sets

$$P := A + \mathcal{D}(A) \quad \text{and} \quad \tilde{P} := A + (\mathcal{D}(A) \setminus \{0_Y\}) \subset P.$$

We associate the dual set

$$T := Y \setminus \tilde{P}$$

with the primal set A.

Example 8.10. We consider the set A and the ordering map $\mathcal{D}$ as specified in Example 2.10, see also Example 8.1. The sets $\mathcal{D}(A)$, P, $\tilde{P}$ and T (see Fig. 8.3) are

$$\mathcal{D}(A) = \{(y_1, y_2) \in \mathbb{R}^2 \mid y_1 \geq 0, \ y_1 + y_2 \geq 0\},$$

$$P = \{(y_1, y_2) \in \mathbb{R}^2 \mid y_1 \in [-1, -1/\sqrt{2}], \ y_2 \geq -\sqrt{1 - y_1^2}\}$$
$$\cup \{(y_1, y_2) \in \mathbb{R}^2 \mid y_1 > -1/\sqrt{2}, \ y_2 \geq -\sqrt{2} - y_1\},$$

$$\tilde{P} = \{(y_1, y_2) \in \mathbb{R}^2 \mid y_1 \in [-1, -1/\sqrt{2}], \ y_2 > -\sqrt{1 - y_1^2}\}$$
$$\cup \{(y_1, y_2) \in \mathbb{R}^2 \mid y_1 > -1/\sqrt{2}, \ y_2 \geq -\sqrt{2} - y_1\},$$

$$T = \{(y_1, y_2) \in \mathbb{R}^2 \mid y_1 < -1\}$$
$$\cup \{(y_1, y_2) \in \mathbb{R}^2 \mid y_1 \in [-1, -1/\sqrt{2}], \ y_2 \leq -\sqrt{1 - y_1^2}\}$$
$$\cup \{(y_1, y_2) \in \mathbb{R}^2 \mid y_1 > -1/\sqrt{2}, \ y_2 < -\sqrt{2} - y_1\}.$$

We briefly state the corresponding results, which can be proofed similarly.

Lemma 8.11. *(i) For the dual set $H := T$ the weak duality (8.2) is satisfied. If $\mathcal{D}(y) \subset \mathcal{D}(A)$ for all $y \in Y$ then also the weak duality (8.1) is satisfied.*
(ii) If $\bar{y} \in A \cap T$, then $\bar{y}$ is a minimal and a nondominated element of A w.r.t. $\mathcal{D}$ and $\bar{y}$ is a maximal element of T w.r.t. $\mathcal{D}$.
If additionally $\mathcal{D}(y) \subset \mathcal{D}(A)$ for all $y \in Y$, then $\bar{y}$ is also a max-nondominated element of T w.r.t. $\mathcal{D}$.

Fig. 8.3 Set T of
Example 8.10, cf. [46]

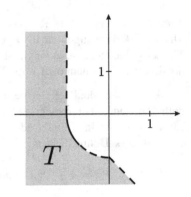

Proof. (i) First we assume that there exists $\bar{y} \in A$ with

$$(\{\bar{y}\} + (\mathcal{D}(\bar{y}) \setminus \{0_Y\})) \cap T \neq \emptyset.$$

However,

$$(\{\bar{y}\} + (\mathcal{D}(\bar{y}) \setminus \{0_Y\})) \subset A + \mathcal{D}(A) \setminus \{0_Y\} = \tilde{P},$$

i.e. $\tilde{P} \cap T \neq \emptyset$, which is a contradiction. Thus (8.2) is satisfied.

Next let $\mathcal{D}(y) \subset \mathcal{D}(A)$ for all $y \in Y$ but there exists an element $\bar{y} \in A$ with $\bar{y} \in \{y\} - (\mathcal{D}(y) \setminus \{0_Y\})$ for some $y \in T$. Then because of $\mathcal{D}(y) \subset \mathcal{D}(A)$ it holds

$$y \in \{\bar{y}\} + (\mathcal{D}(y) \setminus \{0_Y\}) \subset A + (\mathcal{D}(A) \setminus \{0_Y\}) = \tilde{P}$$

in contradiction to $y \in T = Y \setminus \tilde{P}$. Thus (8.1) is satisfied.

(ii) As $\bar{y} \notin \tilde{P}$ we conclude $(\{\bar{y}\} - (\mathcal{D}(A) \setminus \{0_Y\})) \cap A = \emptyset$ and thus

$$(\{\bar{y}\} - (\mathcal{D}(\bar{y}) \setminus \{0_Y\})) \cap A = \emptyset.$$

Hence $\bar{y}$ is a minimal element of A w.r.t. $\mathcal{D}$.

Further, $\bar{y} \notin \{y\} + (\mathcal{D}(A) \setminus \{0_Y\})$ for all $y \in A$ and so $\bar{y} \notin \{y\} + (\mathcal{D}(y) \setminus \{0_Y\})$ for all $y \in A$. Thus $\bar{y}$ is a nondominated element of A w.r.t. $\mathcal{D}$.

Since

$$\emptyset = T \cap \tilde{P} = T \cap (A + \mathcal{D}(A) \setminus \{0_Y\})$$

we conclude $T \cap (\{\bar{y}\} + \mathcal{D}(\bar{y})) = \{\bar{y}\}$. Therefore, $\bar{y}$ is also a maximal element of T w.r.t. $\mathcal{D}$.

Finally, let $\mathcal{D}(y) \subset \mathcal{D}(A)$ for all $y \in Y$ but $\bar{y}$ is not a max-nondominated element of T w.r.t. $\mathcal{D}$. Then there exists $y \in T = Y \setminus \tilde{P}$ with $\bar{y} \in \{y\} - (\mathcal{D}(y) \setminus \{0_Y\})$. We obtain

$$y \in \{\bar{y}\} + (\mathcal{D}(y) \setminus \{0_Y\}) \subset A + (\mathcal{D}(A) \setminus \{0_Y\}) = \tilde{P}$$

in contradiction to $y \in Y \setminus \tilde{P}$.   □

$\mathcal{D}(y) \subset \mathcal{D}(A)$ for all $y \in Y$ is not such a strong assumption as for modeling an application problem one is in general only interested in the cones $\mathcal{D}(y)$ for $y \in A$. We arrive at the following converse duality theorem:

Theorem 8.12. *If $Y \setminus P$ is algebraically open, then every maximal element $\bar{y}$ of T w.r.t. $\mathcal{D}$ is also a minimal element of A w.r.t. $\mathcal{D}$.*

Proof. Let $\bar{y}$ be an arbitrary maximal element of T w.r.t. $\mathcal{D}$, i.e.

$$(\{\bar{y}\} + \mathcal{D}(\bar{y})) \cap T = \{\bar{y}\}.$$

We assume $\bar{y} \notin P$. As $Y \setminus P$ is algebraically open, there exists for every $d \in \mathcal{D}(\bar{y}) \setminus \{0_Y\}$ a scalar $\lambda > 0$ with

$$\bar{y} + \lambda d \in Y \setminus P \subset T.$$

As

$$\bar{y} + \lambda d \in \{\bar{y}\} + (\mathcal{D}(\bar{y}) \setminus \{0_Y\})$$

this is a contradiction to the maximality of $\bar{y}$. Thus $\bar{y} \in P = A + \mathcal{D}(A)$ and due to $\bar{y} \notin \tilde{P} = A + (\mathcal{D}(A) \setminus \{0_Y\})$ we conclude $\bar{y} \in A$. Together with Lemma 8.11(ii) we get that $\bar{y}$ is also a minimal element of A w.r.t. $\mathcal{D}$.   □

Remark 8.13. Note, that under the assumptions of Theorem 8.12 together with the assumptions of Lemma 8.11(ii), $\bar{y}$ is also a nondominated element of A w.r.t. $\mathcal{D}$ and a max-nondominated element of T w.r.t. $\mathcal{D}$.

Example 8.14. We consider the set A and the ordering map $\mathcal{D}$ as specified in Example 2.10; see also Examples 8.1 and 8.10. Here, $Y \setminus P$ is algebraically open. The maximal elements of the set T w.r.t. the ordering map $\mathcal{D}$ are

$$\left\{(y_1, y_2) \in \mathbb{R}^2 \,\middle|\, y_1 \in [-1, -1/\sqrt{2}], \; y_2 = -\sqrt{1 - y_1^2}\right\}.$$

These elements are also minimal elements of A w.r.t. $\mathcal{D}$.

The set $Y \setminus P$ is for instance algebraically open if P is convex and algebraically closed [94, p. 191]. But in general we need no convexity assumptions in Theorem 8.12.

A standard non-converse duality theorem can only be stated under strong assumptions, as a minimal element of the set A (and also a nondominated element) w.r.t. $\mathcal{D}$ does in general not have to be included in the set T:

Fig. 8.4 (a) Sets A and T
of Example 8.15. (b) Sets
A and T of Example 8.17,
cf. [46]

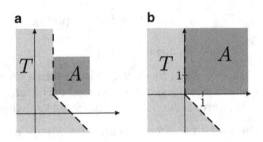

Example 8.15. Let A and $\mathcal{D}$ be specified as in Example 8.8. The set of minimal elements of A w.r.t. the ordering map $\mathcal{D}$ is

$$\{(y_1, y_2) \in \mathbb{R}^2 \mid y_1 = 1, \ y_2 \in [1, 3]\}$$

which equals the set of nondominated elements of A w.r.t. $\mathcal{D}$, while the dual set is

$$T = \{(y_1, y_2) \in \mathbb{R}^2 \mid y_1 < 1\} \cup \{(y_1, y_2) \in \mathbb{R}^2 \mid y_1 \geq 1, \ y_1 + y_2 < 2\}$$
$$\cup \{(1, 1)\};$$

see Fig. 8.4a. The point $(1, 1)$ is maximal for T and minimal for A w.r.t. $\mathcal{D}$, but all the other minimal (and nondominated) elements of A w.r.t. $\mathcal{D}$ are not included in T.

Lemma 8.16. *Let $\bar{y} \in A$ be a minimal element of A w.r.t. $\mathcal{D}$. Then $\bar{y}$ is a maximal element of T w.r.t. $\mathcal{D}$ if and only if*

$$(\{\bar{y}\} - \mathcal{D}(A)) \cap A = \{\bar{y}\},$$

i.e., in the case of $\mathcal{D}(A)$ a pointed convex cone, if and only if $\bar{y}$ is an efficient element of A w.r.t. the partial ordering introduced by the cone $K := \mathcal{D}(A)$.

Proof. Let $(\{\bar{y}\} - \mathcal{D}(A)) \cap A = \{\bar{y}\}$. Then $\bar{y} \notin A + \mathcal{D}(A) \setminus \{0_Y\}$, i.e. $\bar{y} \in T$. Thus $\bar{y} \in A \cap T$ and the assertion follows with Lemma 8.11(ii).

Next, let us assume that $\bar{y} \in A$ is a minimal element of A w.r.t. $\mathcal{D}$ and a maximal element of T w.r.t. $\mathcal{D}$, but it does not hold $(\{\bar{y}\} - \mathcal{D}(A)) \cap A = \{\bar{y}\}$. Then an element $y \in A$ exists with $\bar{y} \in \{y\} + (\mathcal{D}(A) \setminus \{0_Y\})$ and thus $\bar{y} \in \tilde{P}$ and $\bar{y} \notin T$. This is a contradiction to the fact that $\bar{y}$ is a maximal element of T w.r.t. $\mathcal{D}$. $\square$

Example 8.17. Let A and $\mathcal{D}$ be specified as in Example 8.9. Then we have $\mathcal{D}(A) = P = \{(y_1, y_2) \in \mathbb{R}^2 \mid y_1 \geq 0, \ y_1 + y_2 \geq 0\}$ and $\tilde{P} = \mathcal{D}(A) \setminus \{(0, 0)\}$. Thus we obtain

$$T = \{(y_1, y_2) \in \mathbb{R}^2 \mid y_1 < 0 \ \vee \ (y_1 \geq 0 \ \wedge \ y_2 < -y_1)\} \cup \{(0, 0)\},$$

see Fig. 8.4b, and $Y \setminus P$ is algebraically open. The unique minimal element of A w.r.t. $\mathcal{D}$ is the point $(0, 0)$, which is also an efficient element w.r.t. $K := \mathcal{D}(A)$, and $(0, 0)$ is also the unique maximal element of T w.r.t. $\mathcal{D}$.

Remark 8.18. Note, that if $\bar{y}$ is an efficient element of A w.r.t. the pointed convex cone $K := \mathcal{D}(A)$, then it is also a nondominated element of A w.r.t. $\mathcal{D}$. Thus it is a necessary condition for a minimal element $\bar{y}$ of A w.r.t. $\mathcal{D}$ to be a maximal element of T w.r.t. $\mathcal{D}$, that $\bar{y}$ is also a nondominated element of A w.r.t. $\mathcal{D}$.

Thus while considering the maximal elements of the dual set, only those minimal (and nondominated) elements of the set A w.r.t. $\mathcal{D}$ can be found which are also efficient w.r.t. $\mathcal{D}(A)$ (assuming $\mathcal{D}(A)$ to be pointed and convex). So this second dual set delivers no relation between the two optimality concepts based on $\leq_1$ and $\leq_2$, and only a subset of the minimal elements of the primal set w.r.t. $\mathcal{D}$—the set of those which are also efficient elements w.r.t. $\mathcal{D}(A)$—is represented by maximal elements of the dual set.

8.2   Duality Based on Linear Scalarization

In this section, we consider the constrained vector optimization problem

$$\text{Minimize } f(x) \quad \text{subject to } x \in S \tag{8.3}$$

under the following assumptions:

Assumption 2. *Let $\hat{S} \neq \emptyset$ be a convex subset of a real linear space X. Let the real topological linear space Y be equipped with a variable ordering structure defined by a cone-valued map $\mathcal{D}: Y \to 2^Y$ with $\mathcal{D}(y)$ a pointed convex cone with nonempty topological interior for all $y \in Y$. Let the real topological linear space Z be partially ordered by a nonempty convex cone C_Z. Let $g: \hat{S} \to Z$ be a C_Z-convex map, let $f: \hat{S} \to Y$ be a map, and assume*

$$S := \{x \in \hat{S} \mid g(x) \in -C_Z\} \neq \emptyset. \tag{8.4}$$

Under these assumptions, the set S is convex. For convenience we recall the problem (8.3) with the set S given explicitly as in (8.4):

$$\text{Minimize } f(x) \quad \text{subject to } x \in \hat{S} \text{ and } g(x) \in -C_Z. \tag{CVP'}$$

We associate a dual problem with the problem (CVP') and we present a weak and a strong duality result. We set

$$D_1 := \{y \in Y \mid \exists t \in \mathcal{D}(y)^* \setminus \{0_{Y^*}\} \text{ and } \exists u \in C_Z^* \text{ with the property}$$
$$(t \circ f + u \circ g)(x) \geq t(y) \text{ for all } x \in \hat{S}\}. \tag{8.5}$$

Here, $\mathcal{D}(y)^* = \{y^* \in Y^* \mid y^*(u) \geq 0 \text{ for all } u \in \mathcal{D}(y)\}$ denotes as usual the topological dual cone of $\mathcal{D}(y)$ in the dual space Y^* for all $y \in Y$.

We consider weakly max-nondominated elements of the set D_1 w.r.t. $\mathcal{D}$, i.e. elements $\bar{y} \in D_1$ such that there exists no $y \in D_1$ with

$$y \in \{\bar{y}\} + \text{int}(\mathcal{D}(y)).$$

Note, that for $\mathcal{D}(y)$ convex and $\text{int}(\mathcal{D}(y)) \neq \emptyset$, we have $\text{int}(\mathcal{D}(y)) = \text{cor}(\mathcal{D}(y))$.

First, we present a weak duality theorem based on linear scalarizations.

Theorem 8.19. *Let Assumption 2 be satisfied. For every $\bar{y} \in D_1$ a functional $t \in \mathcal{D}(\bar{y})^* \setminus \{0_{Y^*}\}$ exists with*

$$t(\bar{y}) \leq t(y) \text{ for all } y \in f(S). \tag{8.6}$$

Moreover, if $\bar{y} \in D_1 \cap f(S)$, i.e. $\bar{y} = f(\bar{x})$ for some $\bar{x} \in S$, then $\bar{y}$ is a weakly minimal element of $f(S)$ and thus $(\bar{x}, \bar{y})$ is a weakly minimal solution of (CVP') w.r.t. the ordering map $\mathcal{D}$.

Proof. We choose $\bar{y} \in D_1$ arbitrarily. Then some $t \in \mathcal{D}(\bar{y})^* \setminus \{0_{Y^*}\}$ and $u \in C_Z^*$ exist with

$$(t \circ f + u \circ g)(x) \geq t(\bar{y}) \text{ for all } x \in \hat{S}.$$

This inequality implies due to $g(x) \in -C_Z$ for all $x \in S$ that $(t \circ f)(x) \geq t(\bar{y})$ for all $x \in S$, hence (8.6).

Further, if additionally $\bar{y} = f(\bar{x})$ for some $\bar{x} \in S$, then (8.6) and Theorem 4.9(b) imply that $\bar{y}$ is a weakly minimal element of the set $f(S)$ w.r.t. $\mathcal{D}$ and the assertion follows. $\square$

In the convex case, we can also state the following duality theorem again relating the notion of minimal to the notion of nondominated elements w.r.t. a variable ordering.

Theorem 8.20. *Let Assumption 2 be satisfied. Let $\bar{y} = f(\bar{x})$ for some $\bar{x} \in S$ and let $(\bar{x}, \bar{y})$ be a weakly minimal solution of (CVP') w.r.t. $\mathcal{D}$, i.e. $\bar{y}$ is a weakly minimal element of $f(S)$ w.r.t. $\mathcal{D}$. Let f be $\mathcal{D}(\bar{y})$-convex on S and $t \in \mathcal{D}(\bar{y})^* \setminus \{0_{Y^*}\}$ with*

$$t(\bar{y}) \leq t(y) \text{ for all } y \in f(S)$$

be given. Additionally, let the scalar-valued optimization problem

$$\inf_{x \in S}(t \circ f)(x) \tag{8.7}$$

be stable, i.e.

$$\inf_{x \in S}(t \circ f)(x) = \sup_{u \in C_Z^*} \inf_{x \in \hat{S}}(t \circ f + u \circ g)(x)$$

and the problem on the right hand side has at least one solution. Then $\bar{y}$ is also a weakly max-nondominated element of D_1 w.r.t. $\mathcal{D}$.

Proof. Because $\bar{y} \in f(S)$ and $t(\bar{y}) \leq t(y)$ for all $y \in f(S)$, we obtain the inequality $t(f(\bar{x})) \leq t(f(x))$ for all $x \in S$. As f is $\mathcal{D}(\bar{y})$-convex, according to the definition it holds for arbitrary $x, y \in S$ and $\lambda \in [0, 1]$ that

$$\lambda f(x) + (1 - \lambda) f(y) \in \{f(\lambda x + (1 - \lambda)y)\} + \mathcal{D}(\bar{y}).$$

This implies with $t \in \mathcal{D}(\bar{y})^* \setminus \{0_{Y^*}\}$

$$(t \circ f)(\lambda x + (1 - \lambda)y) \leq \lambda(t \circ f)(x) + (1 - \lambda)(t \circ f)(y).$$

Thus the composite map $t \circ f$ is convex on S, and $\bar{x}$ is a minimal solution of the convex optimization problem (8.7), which is assumed to be stable.

As (8.7) is stable, there exists some $\bar{u} \in C_Z^*$ with

$$\inf_{x \in S}(t \circ f)(x) = \inf_{x \in \hat{S}}(t \circ f + \bar{u} \circ g)(x)$$

and thus

$$(t \circ f + \bar{u} \circ g)(\hat{x}) \geq \inf_{x \in S}(t \circ f)(x) = (t \circ f)(\bar{x}) = t(\bar{y}) \text{ for all } \hat{x} \in \hat{S},$$

i.e. $\bar{y} \in D_1$ and hence $\bar{y} \in f(S) \cap D_1$.

Next we show that $\bar{y}$ is a weakly max-nondominated element of D_1, i.e. there is no $y \in D_1$ such that $\bar{y} \in \{y\} - \text{int}(\mathcal{D}(y))$. Indeed, according to Theorem 8.19, for $\bar{y} \in f(S)$ and any $y \in D_1$ a $t \in \mathcal{D}(y)^* \setminus \{0_{Y^*}\}$ exists with $t(y) \leq t(\bar{y})$, i.e. $t(y - \bar{y}) \leq 0$. By Lemma 1.7

$$\text{int}(\mathcal{D}(y)) = \{z \in Y \mid t(z) > 0 \text{ for all } t \in \mathcal{D}(y)^* \setminus \{0_{Y^*}\}\},$$

i.e. $y - \bar{y} \notin \text{int}(\mathcal{D}(y))$. Therefore, $y - \bar{y} \notin \text{int}(\mathcal{D}(y))$ for all $y \in D_1$, as it was to be shown. $\qquad\square$

Remark 8.21. A functional $t \in \mathcal{D}(\bar{y})^* \setminus \{0_{Y^*}\}$ with $t(\bar{y}) \leq t(y)$ for all $y \in f(S)$ exists for a weakly minimal element $\bar{y}$ of $f(S)$ w.r.t. $\mathcal{D}$ according to Theorem 4.9(a) if $f(S)$ is convex.

8.3 Duality Based on Nonlinear Scalarization

In this section we establish again duality results for the constrained vector optimization problem (CVP') (for the definition see page 147) but now we base our examinations on the nonlinear scalarization functionals considered in Chap. 6 for variable ordering structures with images BP cones. Thus, we need the following assumptions.

Assumption 3. *Let $\hat{S}$ be a nonempty subset of a real linear space X. Let the real normed space $(Y, \|\cdot\|)$ be equipped with a variable ordering structure defined by a cone-valued map $\mathcal{D}: Y \to 2^Y$ with*

$$\mathcal{D}(y) = C(\ell(y)) = \{u \in Y \mid \|u\| \le \ell(y)(u)\} \text{ for all } y \in Y$$

for some given map $\ell: Y \to Y^*$. Let the real topological linear space Z be partially ordered by a nonempty convex cone C_Z. Let $g: \hat{S} \to Z$ and $f: \hat{S} \to Y$ be given maps, and assume

$$S = \{x \in \hat{S} \mid g(x) \in -C_Z\} \ne \emptyset.$$

Note that no convexity of $\hat{S}$ or g is assumed. We now use the scalarization functionals presented in Sect. 6.1 which allow a scalar characterization of different kinds of (locally) nondominated solutions of the problem (CVP'). We define the dual set by

$$D_2 := \{\bar{y} \in Y \mid \exists u \in C_Z^* \text{ with } (\xi_{\bar{y}} \circ f + u \circ g)(x) \ge 0 \text{ for all } x \in \hat{S}\}, \quad (8.8)$$

where $\xi_{\bar{y}}$ for some $\bar{y} \in Y$ is, as in Chap. 6, the functional

$$\xi_{\bar{y}}(y) = \ell(y)(y - \bar{y}) + \|y - \bar{y}\| \text{ for all } y \in Y.$$

Recall that $\xi_{\bar{y}}(\bar{y}) = 0$.

We first obtain a weak duality theorem which states that each element of the dual set given above is a "lower bound" for the values of the primal problem (CVP').

Theorem 8.22. *Let the Assumption 3 be satisfied. Let $\bar{y} \in D_2$. Then for any $\hat{y} \in f(S)$ one has*

$$\xi_{\bar{y}}(\hat{y}) \ge \xi_{\bar{y}}(\bar{y}) = 0. \quad (8.9)$$

Moreover, if $\bar{y} \in D_2 \cap f(S)$, i.e. $\bar{y} = f(\bar{x})$ for some $\bar{x} \in S$, and $\|\ell(y)\|_* > 1$ for all $y \in f(S)$ (i.e. $int(\mathcal{D}(y)) \ne \emptyset$), then $\bar{y}$ is a weakly nondominated element of $f(S)$ and thus $(\bar{x}, \bar{y})$ is a weakly nondominated solution of (CVP') w.r.t. the ordering map $\mathcal{D}$.

Proof. By the definition of the set D_2, there exists $u \in C_Z^*$ with

$$(\xi_{\bar{y}} \circ f + u \circ g)(x) \ge 0 \text{ for all } x \in \hat{S}.$$

Suppose that $\hat{y} = f(\hat{x})$ for some $\hat{x} \in S$. Then $g(\hat{x}) \in -C_Z$ and

$$(u \circ g)(\hat{x}) \leq 0$$

and therefore, $(\xi_{\bar{y}} \circ f)(\hat{x}) \geq 0$ or $\xi_{\bar{y}}(\hat{y}) \geq 0$. Thus (8.9) holds. Further, if additionally $\bar{y} = f(\bar{x})$ for some $\bar{x} \in S$ and $\|\ell(y)\|_* > 1$ for all $y \in f(S)$, then (8.9) and Theorem 6.1(c) imply that $\bar{y}$ is a weakly nondominated element of the set $f(S)$ and the assertion follows. $\qquad\square$

Under additional assumptions including the linearity of f and the convexity of S we have a strong duality result.

Theorem 8.23. *Let the Assumption 3 be satisfied and let $\bar{y} = f(\bar{x})$ for some $\bar{x} \in S$. Assume that $\hat{S}$ is convex, f is linear, g is C_Z-convex and $\xi_{\bar{y}}$ is convex (in particular, ℓ is linear and monotone) and $\|\ell(y)\|_* > 1$ for all $y \in f(S)$. Assume further that the scalar optimization problem*

$$\inf_{x \in S} \xi_{\bar{y}}(f(x)) \tag{8.10}$$

is stable, i.e.

$$\inf_{x \in S}(\xi_{\bar{y}} \circ f)(x) = \sup_{u \in C_Z^*} \inf_{x \in \hat{S}}(\xi_{\bar{y}} \circ f + u \circ g)(x)$$

and the problem on the right hand side has at least one solution. If $(\bar{x}, \bar{y})$ is a weakly nondominated solution of (CVP') w.r.t. $\mathcal{D}$, i.e. $\bar{y}$ is a weakly nondominated element of $f(S)$ w.r.t. $\mathcal{D}$, then $\bar{y}$ is also a weakly maximal element of D_2 w.r.t. $\mathcal{D}$.

Proof. Note that under these assumptions the set S is convex. Since the functional $\xi_{\bar{y}}$ is convex (by the assumption or by the linearity and monotonicity of ℓ, according to Lemma 6.7), one can easily prove that the composite map $\xi_{\bar{y}} \circ f$ is convex. Since $\bar{y}$ is a weakly nondominated element of $f(S)$, Theorem 6.1(c), see also Corollary 7.1(d), implies

$$\xi_{\bar{y}}(f(\bar{x})) = 0 \leq \xi_{\bar{y}}(f(x)) \quad \text{for all } x \in S.$$

Hence, $\bar{x}$ is a minimal solution of the convex optimization problem (8.10), which is assumed to be stable. Therefore, one can find $\bar{u} \in C_Z^*$ with

$$\inf_{x \in S}(\xi_{\bar{y}} \circ f)(x) = \inf_{x \in \hat{S}}(\xi_{\bar{y}} \circ f + \bar{u} \circ g)(x)$$

and thus

$$(\xi_{\bar{y}} \circ f + \bar{u} \circ g)(\hat{x}) \geq \inf_{x \in S}(\xi_{\bar{y}} \circ f)(x) = \xi_{\bar{y}}(\bar{y}) = 0 \quad \text{for all } \hat{x} \in \hat{S},$$

i.e. $\bar{y} \in D_2$ and hence $\bar{y} \in f(S) \cap D_2$.

Next we show that $\bar{y}$ is a weakly maximal element of D_2 w.r.t. $\mathcal{D}$, i.e., there is no $y \in D_2$ such that $y - \bar{y} \in \text{int}(\mathcal{D}(\bar{y}))$. Indeed, according to Theorem 8.22, for $\bar{y} \in f(S)$ and any $y \in D_2$ one has $\xi_y(\bar{y}) \geq 0$, i.e.

$$\|y - \bar{y}\| \geq \ell(\bar{y})(y - \bar{y}),$$

which implies with Lemma 1.16(iv) that $y - \bar{y} \notin \text{int}(\mathcal{D}(\bar{y}))$. Therefore, $y - \bar{y} \notin \text{int}(\mathcal{D}(\bar{y}))$ for all $y \in D_2$, as it was to be shown.   □

Note that in contrast to the Theorems 8.22 and 8.23, the duality results in the Theorems 8.19 and 8.20 involve linear scalarization functionals and they concern weakly minimal solutions of the primal problem (CVP') w.r.t. $\mathcal{D}$ and these solutions are related to weakly max-nondominated elements of the dual set. Moreover, note that under the assumptions of Theorem 8.23 the set S is convex and as f is linear, also $f(S)$ is convex. Nevertheless, we use a nonlinear scalarization functional here.

8.4   Notes on the Literature

The weak duality, also denoted dominance requirement, as introduced here, is taken from Gerstewitz (Tammer) [64]. For the strong duality we refer to the book by Göpfert et al. [71, p. 155]. The presentation of the general duality and the duality results based on linear scalarizations is based on [46]. Duality based on nonlinear scalarizations as presented in Sect. 8.3 is due to Eichfelder and Ha [52]. The proof of Theorem 8.6 follows the proof of Theorem 8.3 in the book by Jahn [94] which gives a duality result in a partially ordered space. Also, the proofs of the Theorems 8.20 and 8.23 use ideas of the proof of Theorem 8.7 in [94]. Duality results for vector optimization problems in a partially ordered space are given for instance in Chap. 8 of [94] and in the book by Boţ et al. [25].

Chapter 9
Numerical Methods

This chapter is on numerical approaches for solving several types of vector optimization problems with a variable ordering structure. We focus on the one hand on discrete vector optimization problems with a finite image set and on the other hand on continuous nonlinear vector optimization problems with a possibly non-convex image set for which we assume that there exists a nontrivial convex cone K with $K \subset \mathcal{D}(y)$ for all y in the image set.

9.1 Discrete Problems

In this section we present a method for determining the nondominated and the minimal elements w.r.t. a variable ordering structure of a finite (image) set A. If the set contains a large number of elements, a pairwise comparison and evaluation of the relation $\leq$ defined by the ordering map $\mathcal{D}$ for all elements may take a long time. For reducing the numerical effort for such problems, for the special case $Y = \mathbb{R}^m$ partially ordered by the natural ordering, i.e. $K = \mathbb{R}^m_+$, a procedure was already given by Jahn in [96], see also [94, Sect. 12.4], based on a procedure firstly presented by Younes in [156] and an algorithmic conception by Graef [94, p. 349], called Graef–Younes algorithm with backward iteration (or Jahn–Graef–Younes algorithm). We present here a procedure which generalizes the method presented there to arbitrary real linear spaces equipped with a variable ordering structure.

As before let Y be a real linear space equipped with a variable ordering structure defined by an ordering map $\mathcal{D}: Y \to 2^Y$ with $\mathcal{D}(y)$ a pointed convex cone for all $y \in Y$. Additionally, let A be a nonempty finite subset of Y. The set A may be, for instance, the image of a vector optimization problem $\min_{x \in S} F(x)$, with $F: S \to Y$ and S some nonempty set, i.e. $F(S) = A$.

G. Eichfelder, *Variable Ordering Structures in Vector Optimization*, Vector Optimization, 153
DOI 10.1007/978-3-642-54283-1_9, © Springer-Verlag Berlin Heidelberg 2014

Algorithm 1 Jahn–Graef–Younes method for nondominated elements

---

Require: $A = \{y^1, \ldots, y^k\}$, $\mathcal{D}(y)$ for all $y \in A$
 1: **(forward iteration)** put $U = \{y^1\}$ and $i = 1$
 2: **while** $i < k$ **do**
 3:       replace i by $i + 1$
 4:       **if** $y^i \notin \{y\} + \mathcal{D}(y)$ for all $y \in U$ **then**
 5:              replace U by $U \cup \{y^i\}$
 6:       **end if**
 7: **end while**
 8: put $\{u^1, \ldots, u^p\} = U$
 9: **(backward iteration)** put $T = \{u^p\}$ and $i = p$
10: **while** $i > 1$ **do**
11:       replace i by $i - 1$
12:       **if** $u^i \notin \{u\} + \mathcal{D}(u)$ for all $u \in T$ **then**
13:              replace T by $\{u^i\} \cup T$
14:       **end if**
15: **end while**
16: put $\{t^1, \ldots, t^q\} = T$
17: **(final comparison)** put $V = \emptyset$ and $i = 0$
18: **while** $i < q$ **do**
19:       replace i by $i + 1$
20:       **if** $t^i \notin \{y\} + \mathcal{D}(y)$ for all $y \in A \setminus T$ **then**
21:              replace V by $V \cup \{y^i\}$
22:       **end if**
23: **end while**
24: **return** set $V \subset A$

9.1.1 Determining Nondominated Elements

We start by presenting the extended Jahn–Graef–Younes algorithm for the determination of the nondominated elements of the discrete, finite set $A =: \{y^1, \ldots, y^k\}$ in Algorithm 1. The result of the algorithm is discussed in the following theorem. It states that the procedure is well-defined and delivers exactly the set of all nondominated elements of A w.r.t. $\mathcal{D}$. Note that the original Jahn–Graef–Younes method for partially ordered spaces [96] consists only of the first (called forward iteration) and the second (called backward iteration) while-loop, while for variable ordering structures which are not transitive and not antisymmetric the third while-loop (final comparison for selected elements) has to be added.

In the original Jahn–Graef–Younes algorithm and also in the algorithm here, we assume that the points of the sets are given as a list with some first and some last element. In the first while-loop (forward iteration), the new set U is built by going through the original set A from the beginning to the end and by comparing each element only with the elements considered so far which have not yet turned out to be dominated. In the second while-loop (backward iteration), the new set T is built similarly but this time by going through the set U and by starting with the last element of the set U. For all elements in the set T, finally a complete comparison with all other elements of the set A has to be done in a final while-loop (final comparison).

Theorem 9.1. *Let U, T and V denote the sets gained by Algorithm 1.*

(a) If $\bar{y}$ is a nondominated element of A w.r.t. $\mathcal{D}$, then $\bar{y} \in U$ and $\bar{y} \in T$.
(b) The elements of the set $T \subset A$ are all nondominated elements of T w.r.t. $\mathcal{D}$.
(c) The set V is exactly the set of all nondominated elements of A w.r.t. $\mathcal{D}$.

Proof. (a) Assume $\bar{y}$ is a nondominated element of A w.r.t. $\mathcal{D}$ but is not in U. Then there exists some $y \in U \subset A$, $y \neq \bar{y}$ with $\bar{y} \in \{y\} + \mathcal{D}(y)$ in contradiction to $\bar{y}$ a nondominated element of the set A w.r.t. $\mathcal{D}$. Next, assume $\bar{y}$ is a nondominated element of A w.r.t. $\mathcal{D}$ but is not in T. According to the first part of the proof, $\bar{y} \in U$. Thus there exists some $y \in T \subset A$, $y \neq \bar{y}$ with $\bar{y} \in \{y\} + \mathcal{D}(y)$ in contradiction to $\bar{y}$ a nondominated element of the set A w.r.t. $\mathcal{D}$.

(b) Let $T =: \{t^1, \ldots, t^q\}$ with $q \leq p \leq k$ and $t^j \in T$ arbitrarily chosen with $1 \leq j \leq q$. We assume the elements of the sets to be ordered in the way they are generated in the algorithm. According to the first while-loop, $t^j \notin \{t^i\} + \mathcal{D}(t^i)$ for all $1 \leq i < j$, and according to the second while-loop, $t^j \notin \{t^i\} + \mathcal{D}(t^i)$ for all $j < i \leq q$. Hence, t^j is a nondominated element of T w.r.t. $\mathcal{D}$.

(c) This is a direct consequence of (a), (b) and the definition of nondominated elements. $\qquad \square$

In case the binary relation $\leq_1$ is transitive and antisymmetric, the algorithm can be stopped after the second while-loop as the set T consists already of all nondominated elements of A w.r.t. $\mathcal{D}$ and thus $T = V$. For the proof of this result we need the following lemma which is related to the external stability of the set of nondominated elements.

Lemma 9.2. *Let the binary relation $\leq_1$ as defined in (1.1) be transitive and antisymmetric and let A be a finite subset of Y. Then for all $y \in A$ there exists a nondominated element $\bar{y}$ of A w.r.t. $\mathcal{D}$ with $y \in \{\bar{y}\} + \mathcal{D}(\bar{y})$.*

Proof. Let $y \in A$ be arbitrarily chosen. If y is a nondominated element of A w.r.t. $\mathcal{D}$ then the assertion is proven. Now, let y be not a nondominated element of A w.r.t. $\mathcal{D}$, i.e. there exists some $y^1 \in A$ with $y^1 \leq_1 y$, $y^1 \neq y$.

If y^1 is nondominated we are done. Otherwise there is some $y^2 \neq y^1$ with $y^2 \leq_1 y^1$ and by the transitivity also $y^2 \leq_1 y$, $y^2 \neq y$. If y^2 is not a nondominated element we can find $y^3 \in A \setminus \{y, y^1, y^2\}$ with $y^3 \leq_1 y^2 \leq_1 y^1 \leq_1 y$ and so on. As A is finite and $\leq_1$ is antisymmetric, this procedure stops with a nondominated element $\bar{y} \in A$ of A w.r.t. $\mathcal{D}$ with $\bar{y} \leq_1 y$. $\qquad \square$

Theorem 9.3. *Let the binary relation $\leq_1$ as defined in (1.1) be transitive and antisymmetric and let A be a finite subset of Y. The nondominated elements of A w.r.t. $\mathcal{D}$ are exactly the elements of the set T gained by the Algorithm 1, i.e. the last while loop can be omitted.*

Proof. According to Theorem 9.1 all nondominated elements of the set A w.r.t. $\mathcal{D}$ are an element of T and all the elements of T are nondominated elements of T w.r.t.

$\mathcal{D}$. It remains to be shown that the elements of T are also nondominated elements of A w.r.t. $\mathcal{D}$. Let $y \in T$ and y be not a nondominated element of A w.r.t. $\mathcal{D}$. Then, according to Lemma 9.2, there exists a nondominated element $\bar{y}$ of A w.r.t. $\mathcal{D}$ with $y \in \{\bar{y}\} + \mathcal{D}(\bar{y}) \setminus \{0_Y\}$. According to Theorem 9.1(b), $\bar{y} \in T$ in contradiction to y a nondominated element of T w.r.t. $\mathcal{D}$.   □

Conditions ensuring the transitivity and antisymmetry of the binary relation $\leq_1$ are given in Lemma 1.10. If the linear space is partially ordered by some pointed convex cone K, the binary relation $\leq_K$ is transitive and antisymmetric and thus in this case the algorithm can be stopped after the second while loop. Note that for $\mathcal{D}(y) \equiv \mathbb{R}_+^m$, i.e. for the naturally ordered case in the Euclidean space $Y = \mathbb{R}^m$, the algorithm presented in [96] is a special case of the algorithm above.

However, without transitivity, the set T does in general not consist of exactly the nondominated elements of the set A w.r.t. $\mathcal{D}$:

Example 9.4. Let $A := \{(0,0), (1,0), (0,2)\}$ and

$$\mathcal{D}(0,0) = \text{cone conv}\{(1,-1), (1,1)\},$$
$$\mathcal{D}(1,0) = \text{cone conv}\{(-1,1), (1,1)\},$$
$$\mathcal{D}(0,2) = \mathbb{R}_+^2 .$$

The unique nondominated element of A w.r.t. the ordering map $\mathcal{D}$ is $(0,0)$ but Algorithm 1 delivers the sets $T = U = \{(0,0), (0,2)\}$. Finally, $V = \{(0,0)\}$.

Already the first while-loop reduces the set A in most cases significantly. In [94, Example 12.19] an example—originally given in [156]—is recalled where some subset A of the naturally ordered Euclidean space with 500,000 points is by the first while-loop reduced to 1,001 points. The number of efficient elements w.r.t. the natural ordering is 471 in that example. In another example, 5,000,000 points are reduced to 3,067 by the first while-loop. There are 1,497 efficient elements in that case. The number of necessary comparisons is thereby reduced because each element has to be compared in line 4 (and in line 12) of the algorithm only with the elements of the set U (the set T, respectively), which have in general much less elements than A.

Such a reduction of the necessary comparisons and by that the necessary evaluations of the binary relation is especially important if the number of elements in the finite set A is very large or if the evaluation of the binary relation is costly. For instance, in [51] the authors considered a problem in medical engineering which led to the vector optimization problem of determining the efficient elements of a finite subset in the space of Hermitian matrices. The finite set A consisted there of around 300,000 matrices and as the partial ordering was given by the cone of positive semidefinite matrices, for each comparison of two elements the eigenvalues of the difference of the two matrices had to be determined.

A discrete set A may also be gained by a discretization of the image set of a continuous vector optimization problem. An example is given in the following, cf. [49].

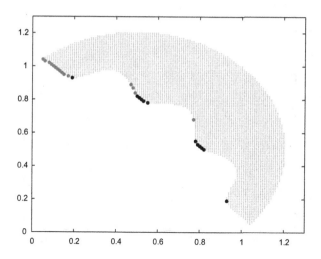

Fig. 9.1 The sets D, U and T of Example 9.5 in *light gray*, *dark gray* and *black*, respectively, cf.
[49]

Example 9.5. We consider the set $A \subset \mathbb{R}^2$ as defined by Tanaka in [144]

$$A = \{x \in [0, \pi] \times [0, \pi] \mid x_1^2 + x_2^2 - 1 - 0.1 \cos \left(16 \arctan(\tfrac{x_1}{x_2}) \right) \geq 0,$$
$$(x_1 - 0.5)^2 + (x_2 - 0.5)^2 \leq 0.5\}.$$

This set might be seen as the image set of the following vector optimization problem:

$$\text{Minimize } f(x) = \begin{pmatrix} x_1 \\ x_2 \end{pmatrix} \text{ subject to } x \in S$$

with $S := A$, the vector-valued objective map $f : \mathbb{R}^2 \to \mathbb{R}^2$ and $f(S) = A$.
We generate a discrete approximation D of this set with 5,014 points by

$$D := A \cap \{(x_1, x_2) \in \mathbb{R}^2 \mid x_1 \in \{0, 0.01, 0.02, \ldots, \pi\},$$
$$x_2 \in \{0.01, 0.02, \ldots, \pi\}\},$$

compare the set of dots in Fig. 9.1.
For the variable ordering structure we define the ordering map $\mathcal{D} : \mathbb{R}^2 \to 2^{\mathbb{R}^2}$ by

$$\mathcal{D}(y) := \{u \in \mathbb{R}^2 \mid \|u\|_2 \leq \ell(y)^\top u\}$$

with

$$\ell(y) := \frac{2}{\min_{i=1,2} y_i} y \quad \text{for all } y \in A,$$

compare (1.13) in Example 1.27 with $\gamma = 1/2$ and $p = (0, 0)$.

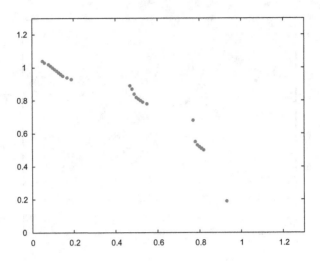

Fig. 9.2 Set U of Example 9.5

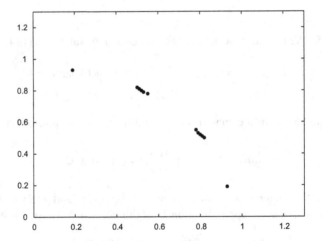

Fig. 9.3 Set T of Example 9.5, cf. [49]

The first while-loop of Algorithm 1 selects 27 points (the set U, compare Fig. 9.2) of the set D as candidates for being nondominated. For that, 61,128 evaluations of the binary relation defined by $\mathcal{D}$ have been necessary.

The second while-loop reduces these 27 points to 12 points, the set T, compare Fig. 9.3, by only 222 additional evaluations of the binary relation. By comparing these remaining points with all other 5,014 points of the discretization (additionally, 60,156 evaluations of the binary relation in the third while-loop) verifies that these 12 points are exactly the nondominated elements of the set D w.r.t. $\mathcal{D}$.

Algorithm 2 Jahn–Graef–Younes method for minimal elements

Require: $A = \{y^1, \ldots, y^k\}$, $\mathcal{D}(y)$ for all $y \in A$
1: **(forward iteration)** put $U = \{y^1\}$ and $i = 1$
2: **while** $i < k$ **do**
3:      replace i by $i + 1$
4:      **if** $y^i \notin \{y\} + \mathcal{D}(y^i)$ for all $y \in U$ **then**
5:          replace U by $U \cup \{y^i\}$
6:      **end if**
7: **end while**
8: put $\{u^1, \ldots, u^p\} = U$
9: **(backward iteration)** put $T = \{u^p\}$ and $i = p$
10: **while** $i > 1$ **do**
11:      replace i by $i - 1$
12:      **if** $u^i \notin \{u\} + \mathcal{D}(u^i)$ for all $u \in T$ **then**
13:          replace T by $\{u^i\} \cup T$
14:      **end if**
15: **end while**
16: put $\{t^1, \ldots, t^q\} = T$
17: **(final comparison)** put $V = \emptyset$ and $i = 0$
18: **while** $i < q$ **do**
19:      replace i by $i + 1$
20:      **if** $t^i \notin \{y\} + \mathcal{D}(t^i)$ for all $y \in A \setminus T$ **then**
21:          replace V by $V \cup \{y^i\}$
22:      **end if**
23: **end while**
24: **return** set $V \subset A$

A pairwise comparison of all 5,014 points with all other elements (till it is shown that an element is dominated by another point or nondominated w.r.t. all) needs 4,472,290 evaluations of the binary relation.

For another approach than the one presented here one might also use Theorem 2.20, in case the assumptions are satisfied, and the classical Jahn–Graef–Younes method for partially ordered spaces to reduce the numerical effort for determining the nondominated elements w.r.t. a variable ordering structure of a finite set.

9.1.2 Determining Minimal Elements

Next, we present in Algorithm 2 the corresponding algorithm for determining the minimal elements of a finite set w.r.t. a variable ordering structure.

Theorem 9.6. *Let U, T and V denote the sets gained by Algorithm 2.*

(a) If $\bar{y}$ is a minimal element of A w.r.t. $\mathcal{D}$, then $\bar{y} \in U$ and $\bar{y} \in T$.
(b) The elements of the set $T \subset A$ are all minimal elements of T w.r.t. $\mathcal{D}$.
(c) The set V is exactly the set of all minimal elements of A w.r.t. $\mathcal{D}$.

Proof. Analogously to the proof of Theorem 9.1.   □

Again, as the following example shows, the elements of T are in general not all also minimal elements of A w.r.t. $\mathcal{D}$. So finally, among all elements of the set T the minimal elements of the set A w.r.t. $\mathcal{D}$ have to be selected in the last while loop (final comparison). For that, for each element $\bar{y}$ of the set T it has to be checked whether there is some $y \in A \setminus T$ with $\bar{y} \in \{y\} + \mathcal{D}(\bar{y}) \setminus \{0_Y\}$, i.e. with

$$y \in \{\bar{y}\} - \mathcal{D}(\bar{y}) \setminus \{0_Y\} .$$

Otherwise, $\bar{y}$ is a minimal element of A w.r.t. $\mathcal{D}$. According to Theorem 9.6(b) it satisfies to compare with the elements of A not being in T.

Example 9.7. Let $A := \{(-1, 1), (0, 0), (1, 0)\}$ and

$$\begin{aligned}
\mathcal{D}(-1, 1) &= \mathbb{R}^2_+, \\
\mathcal{D}(0, 0) &= \text{cone conv}\{(1, -1), (1, 1)\}, \\
\mathcal{D}(1, 0) &= \mathbb{R}^2_+.
\end{aligned}$$

The unique minimal element of A w.r.t. $\mathcal{D}$ is $(-1, 1)$ but Algorithm 2 delivers the sets $T = U = \{(-1, 1), (1, 0)\}$.

We reconsider Example 9.5 and determine the minimal elements of the finite set $D \subset A$ with the help of Algorithm 2, cf. [49].

Example 9.8. We reconsider the set $A \subset \mathbb{R}^2$ and the finite set $D \subset A$ as defined in Example 9.5.

(a) For the variable ordering structure we first consider again the ordering map

$$\mathcal{D}(y) := \{u \in \mathbb{R}^2 \mid \|u\|_2 \le \ell(y)^\top u\}$$

with

$$\ell(y) := \frac{2}{\min_{i=1,2} y_i} y \text{ for all } y \in A,$$

compare (1.13) with $\gamma = 1/2$ and $p = (0, 0)$.

The first while-loop of Algorithm 2 selects 18 points of the set D as candidates for being minimal. For that only 7,036 evaluations of the binary relation defined by $\mathcal{D}$ have been necessary. The second while-loop reduces these 18 points to 5 points, the set T, by only 23 additional evaluations of the binary relation. By comparing these remaining points with all other 5,014 points of the discretization (additionally, 22,119 evaluations of the binary relation in the third while-loop) however shows that none of the 5 points of the set T is a minimal element of D w.r.t. $\mathcal{D}$. Thus, there are no minimal elements at all in D w.r.t. the variable ordering structure. A pairwise comparison of all 5,014 points with all

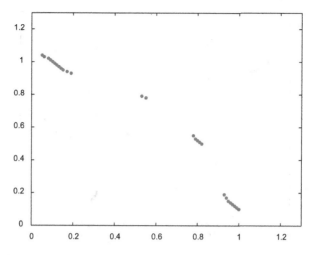

Fig. 9.4 Set U of Example 9.8(b)

other elements (till it is shown that an element is not a minimal element w.r.t.
$\mathcal{D}$) needs 582,538 evaluations of the binary relation.
(b) Next, we consider another variable ordering structure defined by the ordering
map $\mathcal{D}: \mathbb{R}^2 \to 2^{\mathbb{R}^2}$ with

$$\mathcal{D}(y) := \{u \in \mathbb{R}^2 \mid \|u\|_2 \leq \ell(y)^\top u\}$$

and

$$\ell(y) := \frac{2}{\min_{i=1,2}(y_i + 1.2)}\left(y + \begin{pmatrix} 1.2 \\ 1.2 \end{pmatrix}\right) \text{ for all } y \in A,$$

compare (1.13) with $\gamma = 1/2$ and $p = (-1.2, -1.2)$.

The first while-loop selects now 27 points (the set U, compare Fig. 9.4) of the
set D as candidates for being minimal elements of D w.r.t. $\mathcal{D}$ (8,625 evaluations
of the binary relation).

The second while-loop reduces to 20 points (213 evaluation), the set T,
compare Fig. 9.5. By comparing these remaining points with all other 5,014
points of the set D (100,260 evaluations in the third while-loop) verifies that
these 20 points are exactly the minimal elements of the set D w.r.t. $\mathcal{D}$.

For the final comparison 100,260 evaluations of the binary relation have been
necessary, but compared to the number of evaluations needed for a pairwise
comparison of all 5,014 points with all other elements till minimality is shown
or a preferred element is detected (453,994 evaluations) this is still a reduction
with the factor 4. In Fig. 9.6 all elements of the set D together with the
determined minimal elements of D w.r.t. $\mathcal{D}$ and the elements of the set U are
shown.

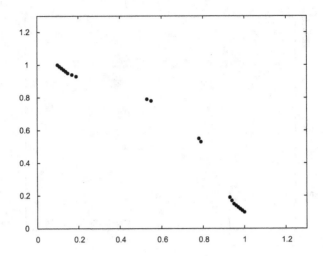

Fig. 9.5 Set T of Example 9.8(b), cf. [49]

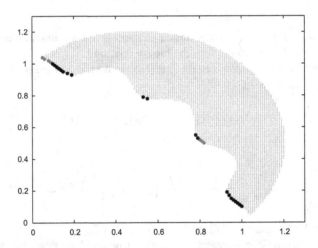

Fig. 9.6 The sets D, U and V of Example 9.8(b) in *light gray*, *dark gray* and *black*, respectively, cf. [49]

The following theorem gives assumptions which ensure that the elements of T are already exactly the minimal elements of A w.r.t. $\mathcal{D}$. We need the following lemma.

Lemma 9.9. *Let the binary relation $\leq_2$ as defined in (1.2) be transitive and antisymmetric and let A be a finite subset of Y. Then for all $y \in A$ there exists a minimal element $\bar{y}$ of A w.r.t. $\mathcal{D}$ with $y \in \{\bar{y}\} + \mathcal{D}(y)$.*

Proof. Let $y \in A$ be arbitrarily chosen. If y is a minimal element of A w.r.t. $\mathcal{D}$ then the assertion is proven. Now, let y be not a minimal element of A w.r.t. $\mathcal{D}$, i.e.

there exists some $y^1 \in A$ with $y^1 \leq_2 y$, $y^1 \neq y$. If y^1 is minimal w.r.t. $\mathcal{D}$ we are done. Otherwise there is some $y^2 \neq y^1$ with $y^2 \leq_2 y^1$ and by the transitivity also $y^2 \leq_2 y$. If y^2 is not a minimal element of A w.r.t. $\mathcal{D}$ we can find $y^3 \in A \setminus \{y, y^1, y^2\}$ with $y^3 \leq_2 y^2 \leq_2 y^1 \leq_2 y$ and so on. As A is finite and $\leq_2$ is antisymmetric, this procedure stops with a minimal element $\bar{y} \in A$ of A w.r.t. $\mathcal{D}$ with $y \in \{\bar{y}\} + \mathcal{D}(y)$. □

Theorem 9.10. *Let the binary relation $\leq_2$ as defined in (1.2) be transitive and antisymmetric and let A be a finite subset of Y. The minimal elements of A w.r.t. $\mathcal{D}$ are exactly the elements of the set T gained by the Algorithm 2, i.e. the last while-loop can be omitted.*

Proof. According to Theorem 9.6 all minimal elements of the set A w.r.t. $\mathcal{D}$ are an element of T and all the elements of T are minimal elements of T w.r.t. $\mathcal{D}$. It remains to be shown that the elements of T are also minimal elements of A w.r.t. $\mathcal{D}$. Let $y \in T$ and y be not a minimal element of A w.r.t. $\mathcal{D}$. Then, according to Lemma 9.9, there exists a minimal element $\bar{y}$ of A w.r.t. $\mathcal{D}$ with $y \in \{\bar{y}\} + \mathcal{D}(y) \setminus \{0_Y\}$. According to Theorem 9.6(b), $\bar{y} \in T$ in contradiction to y a minimal element of T w.r.t. $\mathcal{D}$. □

Note that one might also use Theorem 2.22 and the classical Jahn–Graef–Younes method for partially ordered spaces to reduce the numerical effort for determining the minimal elements w.r.t. a variable ordering structure of a finite set.

9.2 Continuous Problems

In this section we present a procedure which allows the determination of an approximation of the set of optimal (nondominated and accordingly minimal) elements of a vector optimization problem with a variable ordering structure where the (image) set A has an infinite number of elements. Such a set A can, for instance, be the image set of a constrained vector optimization problem as discussed in the Examples 7.18 and 9.5. As the set of optimal elements is then in general infinite, only an approximation can be computed in general. This approximation should thereby be of a high quality. Several quality measures exist; for an overview see [42]. They all evaluate mainly whether the whole set of optimal elements is approximated and whether the approximation points are equidistantly distributed.

Note that for determining a single nondominated or minimal element, an efficient element w.r.t. $\mathcal{D}(A)$ can be computed (assuming $\mathcal{D}(A)$ to be a pointed convex cone), compare Lemmas 2.15 and 2.16. However, in the following we are interested in the determination of an approximation of the whole set of optimal (nondominated and accordingly minimal) elements.

As before, let Y be a real topological linear space equipped with a variable ordering structure defined by the ordering map $\mathcal{D}: Y \to 2^Y$ with $\mathcal{D}(y)$ a pointed convex cone and let A be a nonempty subset of Y. Further, let K be a closed pointed convex cone with

$$\{0_Y\} \neq K \subset \bigcap_{y \in A} \mathcal{D}(y).$$

We need two important results which we have already given in Chap. 2 in a slightly modified form:

Lemma 9.11. *(i) Any nondominated element of A w.r.t. $\mathcal{D}$ is also an efficient element of A with the space Y partially ordered by any convex cone K with $K \subset \bigcap_{y \in A} \mathcal{D}(y)$.*

(ii) Any minimal element of A w.r.t. $\mathcal{D}$ is also an efficient element of A with the space Y partially ordered by any convex cone K with $K \subset \bigcap_{y \in A} \mathcal{D}(y)$.

Proof. (i) See Lemma 2.37(i).

(ii) According to Lemma 2.15(i) any minimal element of A w.r.t. $\mathcal{D}$ is also an efficient element of A with the space Y partially ordered by $\mathcal{D}(\bar{y})$ and thus also if partially ordered by K.

$\square$

This lemma delivers a useful necessary condition for determining a subset of A which contains all optimal (minimal and nondominated) elements w.r.t. the variable ordering structure. Note that for instance in [58] ordering maps were proposed in the Euclidean space $\mathbb{R}^m$ for modeling preferences of decision makers with $\mathbb{R}_+^m \subset \mathcal{D}(y)$ for all $y \in A$. Thus in that case we can choose

$$K := \mathbb{R}_+^m \subset \bigcap_{y \in A} \mathcal{D}(y),$$

see Example 1.27.

The set of efficient elements w.r.t. K denoted with $\mathcal{E}_K$ delivers thus a superset of the set of nondominated (and minimal) elements of A w.r.t. $\mathcal{D}$. As in general not the complete set of efficient elements $\mathcal{E}_K$ w.r.t. K can be determined, we generate an approximation. For such an approximation in finite dimensional spaces many methods can be found in the literature. We use here a procedure introduced in [40] for $Y = \mathbb{R}^m$, see also [41–45], which generates an approximation with a high quality of the efficient set of the image set of a vector optimization problem for which it is assumed that, among others, some differentiability assumptions are satisfied. The procedure uses the scalarization functional given in (5.5) in Sect. 5.2, i.e.

$$\psi_{a,r}(y) = \inf\{t \in \mathbb{R} \mid a + t\,r - y \in K\},$$

and determines the parameter a adaptively based on sensitivity information. The method is especially appropriate for determining approximations of the whole efficient set in lower dimensions as for $Y = \mathbb{R}^2$. Thereby it can only be guaranteed that weakly efficient approximation points are determined.

We denote the finite set of approximation points of $\mathcal{E}_K$, determined by a classical approximation method as the one mentioned above which delivers at least weakly efficient elements, by $\mathcal{E}_K^{\mathrm{approx}}$. For selecting from $\mathcal{E}_K^{\mathrm{approx}}$ an approximation of the set of optimal (i.e., nondominated or minimal) elements of A w.r.t. $\mathcal{D}$, first the optimal elements of $\mathcal{E}_K^{\mathrm{approx}}$ w.r.t. $\mathcal{D}$ can be selected. For this selection, Algorithm 1 and accordingly Algorithm 2 can be used. By that, we reduce the set $\mathcal{E}_K^{\mathrm{approx}}$ to some finite subset denoted with $T \subset \mathcal{E}_K^{\mathrm{approx}} \subset A$. Each element of T is thus an at least weakly efficient element of A and according to Theorems 9.1(c) and 9.6(c) a nondominated/minimal element of $\mathcal{E}_K^{\mathrm{approx}}$ w.r.t. $\mathcal{D}$, respectively.

In the following two subsections, we discuss the determination of an approximation of the set of the nondominated and of the minimal elements w.r.t. $\mathcal{D}$, respectively.

9.2.1 Determining Nondominated Elements

With the procedure presented so far, each element of T is an at least weakly efficient element of A w.r.t. K and by Theorem 9.1(c) also a nondominated element of $\mathcal{E}_K^{\mathrm{approx}}$ w.r.t. $\mathcal{D}$. If $\mathcal{E}_K$ is externally stable and assumption (2.7) of Theorem 2.20 is satisfied, then the nondominated elements of A w.r.t. $\mathcal{D}$ are exactly those efficient elements of A w.r.t. K which are nondominated elements of $\mathcal{E}_K$ w.r.t. $\mathcal{D}$. Based on this theorem, in [86] the set T is taken as an approximation of the set of nondominated elements of A w.r.t. $\mathcal{D}$. Thereby, the set T is also obtained by selecting the nondominated elements of $\mathcal{E}_K^{\mathrm{approx}}$ w.r.t. $\mathcal{D}$ while the finite set $\mathcal{E}_K^{\mathrm{approx}}$ is gained by applying a multiobjective evolutionary algorithm.

However, we need that the elements of the set T are nondominated elements of $\mathcal{E}_K$, and not of $\mathcal{E}_K^{\mathrm{approx}}$ only, w.r.t. $\mathcal{D}$. There might be some element $t \in T$ with

$$t \notin \{y\} + \mathcal{D}(y) \setminus \{0_Y\} \quad \text{for all } y \in \mathcal{E}_K^{\mathrm{approx}}$$

but

$$t \in \{y\} + \mathcal{D}(y) \setminus \{0_Y\} \quad \text{for some } y \in \mathcal{E}_K \setminus \mathcal{E}_K^{\mathrm{approx}}. \tag{9.1}$$

For that reason we use the scalarization results as presented in Chaps. 5 and 6 for selecting those points in T which are in fact (weakly) nondominated elements of A w.r.t. $\mathcal{D}$. Of course, if $\mathcal{E}_K^{\mathrm{approx}}$ is an approximation of $\mathcal{E}_K$ with only small distances between the approximation points and thus with a high quality and the cone-valued map $\mathcal{D}$ has some continuous behavior, the situation as described in (9.1) might not appear.

In the following algorithm we use the scalarization result of Theorem 5.11. Let $r \in \mathrm{int} K$. Then $\bar{y}$ is a nondominated element of A w.r.t. the ordering map $\mathcal{D}$ if and only if

$$\inf\{t \in \mathbb{R} \mid \bar{y} + t\,r - y \in \mathcal{D}(y)\} > 0 \text{ for all } y \in A \setminus \{\bar{y}\} \qquad (9.2)$$

and $\bar{y}$ is a weakly nondominated element of A w.r.t. the ordering map $\mathcal{D}$ if and only if

$$\inf\{t \in \mathbb{R} \mid \bar{y} + t\,r - y \in \mathcal{D}(y)\} \geq 0 \text{ for all } y \in A. \qquad (9.3)$$

If we apply a numerical solution method to solve (9.2) we can in general not verify the strict inequality and thus we can only show the weak nondominatedness of some element. If the ordering map has images being representable as BP cones in the normed space $\mathbb{R}^m$, i.e. if

$$\mathcal{D}(y) = \{u \in Y \mid \|u\| \leq \ell(y)^{\mathsf{T}} u\} \text{ for all } y \in \mathbb{R}^m$$

for some map $\ell : \mathbb{R}^m \to \mathbb{R}^m$, then the optimization problem related to (9.3) reads for instance as

Minimize t subject to $\|\bar{y} + t\,r - y\| - \ell(y)^{\mathsf{T}}(\bar{y} + t\,r - y) \leq 0,\ t \in \mathbb{R},\ y \in A.$

Thus, due to the norm in the constraint, the above scalar-valued optimization problem is a nondifferentiable nonlinear optimization problem which may even be discontinuous dependently on ℓ. For numerically solving such problems adequate solution methods are necessary.

We sum up the procedure in the algorithm on page 167.

Theorem 9.12. *Let N denote the set generated by Algorithm 3. Then $N \subset A$ and any element of N is a weakly nondominated element of A w.r.t. $\mathcal{D}$.*

Proof. $N \subset W \subset \mathcal{E}_K^{\mathrm{approx}} \subset A$ and by Step 6 and Theorem 5.11(b) all elements of N are weakly nondominated elements of A w.r.t. $\mathcal{D}$. $\qquad \square$

Example 9.13 ([49]). We consider the set $A \subset \mathbb{R}^2$ as introduced in Example 9.5 which was originally defined in [144]

$$A = \{x \in [0, \pi] \times [0, \pi] \mid x_1^2 + x_2^2 - 1 - 0.1 \cos\left(16 \arctan(\tfrac{x_1}{x_2})\right) \geq 0,$$
$$(x_1 - 0.5)^2 + (x_2 - 0.5)^2 \leq 0.5\}.$$

For the variable ordering structure we define the ordering map again by

$$\mathcal{D}(y) = \{u \in \mathbb{R}^2 \mid \|u\|_2 \leq \ell(y)^{\mathsf{T}} u\}$$

Algorithm 3 Algorithm for the approximation of the set of nondominated elements

---

Require: set $A \subset \mathbb{R}^m$, $\mathcal{D}(y)$ for all $y \in A$, $K \subset \bigcap_{y \in A} \mathcal{D}(y)$ a convex cone and
   $r \in \text{int}(K)$
1: **(approximation)** determine an approximation $\mathcal{E}_K^{\text{approx}} \subset A$ of the set of efficient elements of
   A w.r.t. K
2: **(reduction)** determine the set of all nondominated elements $W = \{y^1, \ldots, y^k\}$ of $\mathcal{E}_K^{\text{approx}}$
   w.r.t. $\mathcal{D}$
3: **(selection)** put $j = 1$ and $N = \emptyset$
4: **while** $j \leq k$ **do**
5:    determine $\bar{t} = \min\{t \in \mathbb{R} \mid y^j + t\,r - y \in \mathcal{D}(y), \ y \in A\}$
6:    **if** $\bar{t} = 0$ **then**
7:       replace N by $N \cup \{y^j\}$
8:    **end if**
9:    replace j by $j + 1$
10: **end while**
11: **return** set $N \subset A$

---

with

$$\ell(y) := \frac{2}{\min_{i=1,2} y_i} y \quad \text{for all } y \in A.$$

Applying Algorithm 3 we first determine an approximation $\mathcal{E}_K^{\text{approx}}$ of the set of efficient elements of A w.r.t. $K := \mathbb{R}_+^2$. According to Example 1.27, $K \subset \mathcal{D}(y)$ for all $y \in A$. The approximation is generated by using the nonlinear scalarization functional as defined in (5.5) and an adaptive parameter control as introduced in [41–43]. This adaptive procedure aims on determining an equidistant approximation of the set of efficient elements of A. As aimed distance we choose 0.02. By this procedure the generated approximation points are guaranteed to be at least weakly efficient elements of A. 55 approximation points are determined shown as dots (all colors) in Fig. 9.7. The curve in Fig. 9.7 shows the boundary of the set A.

Next, Algorithm 1 is applied to the set of approximation points $\mathcal{E}_K^{\text{approx}}$. After the first while loop, 26 points are deleted (marked in light gray in Fig. 9.7) and only 29 points remain. In the second while loop, 19 points are deleted (marked in dark gray) and 10 points remain. For all of these 10 points (marked in black) it is verified in the third while-loop that they are nondominated elements of the set $\mathcal{E}_K^{\text{approx}}$ w.r.t. $\mathcal{D}$. These 10 points equal the set W in the algorithm above.

For all 10 points y^j the optimization problem

$$\min\{t \in \mathbb{R} \mid y^j + t\,r - y \in \mathcal{D}(y), \ y \in A\} \tag{9.4}$$

is solved to check whether they are weakly nondominated elements of A w.r.t. $\mathcal{D}$. This resulted in 6 remaining points (those with zero as minimal value in (9.4)), see Fig. 9.8.

However, note that the scalar-valued optimization problem (9.4) is not everywhere differentiable due to the definition of $\mathcal{D}$ and it is nonconvex and nonlinear,

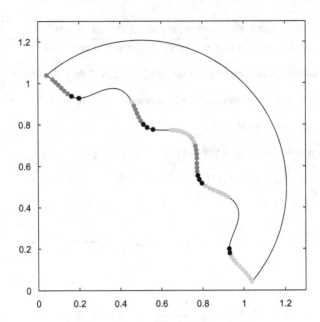

Fig. 9.7 Set $\mathcal{E}_K^{\text{approx}}$ of Example 9.13. The different colors of the dots are explained in the text, cf. [49]

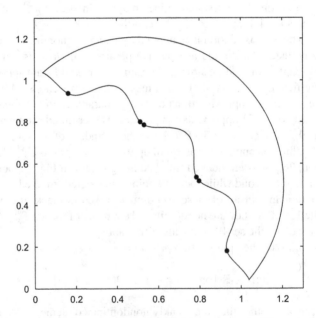

Fig. 9.8 Approximation of the set of nondominated elements w.r.t. $\mathcal{D}$ of Example 9.13, cf. [49]

but for this example we nevertheless just applied a standard numerical solver pre-implemented in Matlab for differentiable optimization problems which does not guarantee to find global minimal solutions for such optimization problems.

The procedure summarized in Algorithm 3 starts by determining an approximation of the set $\mathcal{E}_K$ of efficient elements of A w.r.t. K with $K \subset \bigcap_{y \in A} \mathcal{D}(y)$, i.e. K might be a small cone and thus the set $\mathcal{E}_K$ quite large. If the approximation points are gained by scalarization techniques and a good approximation of $\mathcal{E}_K$ should be gained with small distances between the approximation points, many scalar-valued optimization problems have to be solved. This is a large numerical effort while many of the determined approximation points may not be nondominated w.r.t. the variable ordering structure $\mathcal{D}$.

9.2.2 Determining Minimal Elements

Next we also give the procedure for determining an approximation of the set of minimal elements of some set A w.r.t. a variable ordering structure. Each element of T is an at least weakly efficient element of A w.r.t. K and by Theorem 9.6(c) also a minimal element of $\mathcal{E}_K^{\text{approx}}$ w.r.t. $\mathcal{D}$.

According to Theorem 2.22, if $\mathcal{E}_K$ is externally stable, the minimal elements of A w.r.t. $\mathcal{D}$ are exactly those efficient elements of A w.r.t. K which are minimal elements of $\mathcal{E}_K$ w.r.t. $\mathcal{D}$. Again, also under these assumptions, we would need those elements of T which are not only minimal elements of $\mathcal{E}_K^{\text{approx}}$ w.r.t. $\mathcal{D}$ but which are also minimal elements of $\mathcal{E}_K$ w.r.t. $\mathcal{D}$. As the set $\mathcal{E}_K$ can in general not be computed, but only an approximation of it, we instead use the scalarization results of Chaps. 5 and 6 for selecting those points in T which are in fact minimal elements of A w.r.t. $\mathcal{D}$.

By Theorem 5.13 for $r \in \text{int} K$ it holds that $\bar{y}$ is a minimal element of A w.r.t. the ordering map $\mathcal{D}$ if and only if

$$\inf\{t \in \mathbb{R} \mid \bar{y} + t\,r - y \in \mathcal{D}(\bar{y})\} > 0 \text{ for all } y \in A \setminus \{\bar{y}\}$$

and $\bar{y}$ is a weakly minimal element of A w.r.t. the ordering map $\mathcal{D}$ if and only if

$$\inf\{t \in \mathbb{R} \mid \bar{y} + t\,r - y \in \mathcal{D}(\bar{y})\} \geq 0 \text{ for all } y \in A \setminus \{\bar{y}\}. \tag{9.5}$$

For instance, if the ordering map has images being representable as BP cones in the normed space Y, i.e. if $\mathcal{D}(y) = \{u \in Y \mid \|u\| \leq \ell(y)^\top(u)\}$ for all $y \in \mathbb{R}^m$ for some map $\ell \colon \mathbb{R}^m \to \mathbb{R}^m$, then the optimization problem in (9.5) reads as

Minimize t subject to $\|\bar{y} + t\,r - y\| - \ell(\bar{y})^\top(\bar{y} + t\,r - y) \leq 0,\ t \in \mathbb{R},\ y \in A$.

We give the procedure for determining minimal elements w.r.t. a variable ordering structure in algorithmic form in Algorithm 4.

Algorithm 4 Algorithm for the approximation of the set of minimal elements

---

Require: set $A \subset \mathbb{R}^m$, $\mathcal{D}(y)$ for all $y \in A$, $K \subset \bigcap_{y \in A} \mathcal{D}(y)$ a convex cone and
$\quad\quad r \in \text{int}(K)$
1: **(approximation)** determine an approximation $\mathcal{E}_K^{\text{approx}} \subset A$ of the set of efficient elements of
$\quad\quad A$ w.r.t. K
2: **(reduction)** determine the set of all minimal elements $W = \{y^1, \ldots, y^k\}$ of $\mathcal{E}_K^{\text{approx}}$ w.r.t. $\mathcal{D}$
3: **(selection)** put $j = 1$ and $M = \emptyset$
4: **while** $j \leq k$ **do**
5: $\quad\quad$ determine $\bar{t} = \min\{t \in \mathbb{R} \mid y^j + t\,r - y \in \mathcal{D}(y^j), \ y \in A\}$
6: $\quad\quad$ **if** $\bar{t} = 0$ **then**
7: $\quad\quad\quad\quad$ replace M by $M \cup \{y^j\}$
8: $\quad\quad$ **end if**
9: $\quad\quad$ replace j by $j + 1$
10: **end while**
11: **return** set $M \subset A$

Theorem 9.14. *Let M denote the set generated by Algorithm 4. Then $M \subset A$ and any element of M is a weakly minimal element of A w.r.t. $\mathcal{D}$.*

Proof. $M \subset W \subset \mathcal{E}_K^{\text{approx}} \subset A$ and by Step 6 and Corollary 5.14(b) all elements of M are weakly minimal elements of A w.r.t. the ordering map $\mathcal{D}$. $\quad\quad\quad\square$

Example 9.15 ([49]). We consider again the set $A \subset \mathbb{R}^2$ of Example 9.15 which was originally defined in [144]. For the variable ordering structure we define the ordering map again as in Example 9.8(b) by $\mathcal{D}(y) = \{u \in \mathbb{R}^2 \mid \|u\|_2 \leq \ell(y)^\top u\}$ with

$$\ell(y) := \frac{2}{\min_{i=1,2}(y_i + 1.2)} \left(y + \begin{pmatrix} 1.2 \\ 1.2 \end{pmatrix} \right) \quad \text{for all } y \in A,$$

compare (1.13) with $\gamma = 1/2$ and $p = (-1.2, -1.2)$.

Applying Algorithm 4, we first determine an approximation $\mathcal{E}_K^{\text{approx}}$ of the set of efficient elements of A w.r.t. $K := \mathbb{R}_+^2$. According to Example 1.27, $K \subset \mathcal{D}(y)$ for all $y \in A$. This approximation is determined as described in Example 9.13 and consists of 55 approximation points, see Fig. 9.9.

Next, Algorithm 2 is applied to the set of approximation points $\mathcal{E}_K^{\text{approx}}$. After the first while loop, 33 points are deleted (marked in light gray in Fig. 9.9) and only 22 points remain. In the second while loop, 9 points are deleted (marked in dark gray) and 13 points remain. For all of these 13 points (marked in black) it is verified in the third while-loop that they are minimal elements of the set $\mathcal{E}_K^{\text{approx}}$ w.r.t. $\mathcal{D}$. These 13 points equal the set W in the algorithm above.

For all 13 points y^j the scalar-valued optimization problem

$$\min\{t \in \mathbb{R} \mid y^j + t\,r - y \in \mathcal{D}(y^j), \ y \in A\}$$

is solved to check whether they are weakly minimal elements of A w.r.t. $\mathcal{D}$. This resulted in 12 remaining points, see Fig. 9.10. However, note that the used numerical

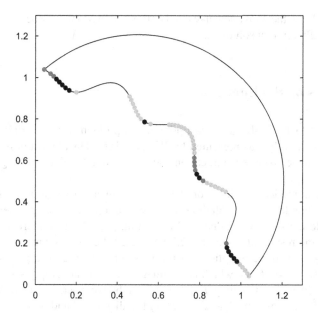

Fig. 9.9 Set $\mathcal{E}_K^{\text{approx}}$ of Example 9.15. The different colors of the dots are explained in the text, cf. [49]

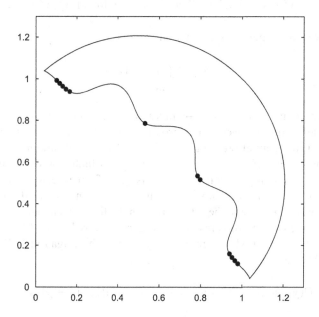

Fig. 9.10 Approximation of the set of nondominated elements of Example 9.15, cf. [49]

solver for solving the above scalar-valued optimization problems does not guarantee
to find globally optimal solutions.

9.3 Notes on the Literature

The first approach to solve a vector optimization problem with a variable ordering
structure numerically was presented by Wacker [146]. His algorithm is especially
designed for the considered application and allows to determine one single optimal
solution.

The first numerical method for determining an approximation of the set of
minimal or nondominated solutions of a vector optimization problem w.r.t. a
variable ordering structure is due to Hirsch et al. [86]. Their numerical procedure
is based on the results of Theorems 2.20 and 2.22. Several assumptions are made
which imply that the assumptions of the mentioned theorems are satisfied: they
assume the set $A := f(S)$, which is the image set of the nonempty set $S \subset \mathbb{R}^n$ by
the vector-valued objective map $f : \mathbb{R}^n \to \mathbb{R}^m$, to be compact. For the ordering map
$\mathcal{D} : \mathbb{R}^m \to 2^{\mathbb{R}^m}$ they assume that $\mathbb{R}^m_+ \subset \mathcal{D}(y)$ for all $y \in \mathbb{R}^m$ and for determining the
nondominated elements they further assume that

$$y^1 - y^2 \in \mathbb{R}^m_+ \text{ implies } \mathcal{D}(y^1) \subset \mathcal{D}(y^2) \text{ for all } y^1, y^2 \in A . \qquad (9.6)$$

Condition (9.6) corresponds to assumption (2.7) as

$$\mathbb{R}^m_+ \subset \bigcap_{y \in A} \mathcal{D}(y).$$

Using evolutionary algorithms for multiobjective optimization problems in a
partially ordered linear space, an approximation $\mathcal{E}_K^{\text{approx}}$ of the set of efficient
elements w.r.t. the natural ordering cone $\mathbb{R}^m_+$ is determined (around 5,000 points
in the test problems), cf. [86, p. 98], and among this finite approximation set the
nondominated (accordingly minimal) elements w.r.t. $\mathcal{D}$ are selected. This selection
is considered to be an approximation of the set of nondominated (accordingly
minimal) elements w.r.t. $\mathcal{D}$ of the set A based on the theorems mentioned above.

In the considered test problems in [86] ordering maps $\mathcal{D}$ with $\mathcal{D}(y) \subset \mathbb{R}^2$ and
$\mathcal{D}(y) \subset \mathbb{R}^3$ are chosen which are of the type given by Engau in [58] (compare
Example 1.27), i.e.

$$\mathcal{D}(y) := \left\{ d \in \mathbb{R}^m \mid d^\top (y - p) \geq \gamma \cdot \|d\|_2 \cdot [y - p]_{\min} \right\} . \qquad (9.7)$$

The parameter γ controls the angle of the cone. They have chosen $\gamma \in \{0.5, 1\}$. For
p they set $p = (0, 0)$, $p = (-1, -1)$ or similar, while $p \in \mathbb{R}^m$ should be an ideal
point, i.e.

$$p_i < \min_{y \in A} y_i \text{ for } i = 1, \ldots, m.$$

It holds $\mathbb{R}^m_+ \subset \mathcal{D}(y)$ and $\mathcal{D}(y)$ is a closed pointed convex cone for all $y \in A$, cf. [58] and Example 1.27. Note that the above cones are Bishop-Phelps cones by setting

$$\ell(y) := \frac{1}{\gamma \, [y - p]_{\min}} (y - p)$$

and

$$\mathcal{D}(y) = C(\ell(y)) := \{u \in Y \mid \|u\|_2 \le \ell(y)^{\top} u\}. \tag{9.8}$$

For testing the algorithm, classical test problems from the literature on multiobjective evolutionary algorithms are taken where the partial ordering is replaced by a variable ordering structure. However, for none of these problems the sets of minimal or nondominated elements w.r.t. the variable ordering structures are known, i.e. the numerical results cannot be verified.

Shukla also generalized this approach together with Braun [136] for set-valued maps $\mathcal{D}: Y \to 2^Y$ with $\mathcal{D}(y)$ a non-convex set under the assumption that there is still some closed pointed nontrivial convex cone K with $K \subset \mathcal{D}(y)$ for all $y \in A$.

In addition to that, also for the notion of equitability a numerical procedure based on evolutionary algorithms has already been presented in the literature. Equitability, defined in $Y = \mathbb{R}^m$, is a stronger concept than efficiency w.r.t. the cone $\mathbb{R}^m_+$, i.e. the equitable efficient elements are a subset of the set of efficient elements w.r.t. the componentwise (natural) partial ordering, see Sect. 1.3.2. This notion corresponds to a variable ordering structure with the sets $\mathcal{D}(y)$ being constant for y within a so-called sector and the space being partitioned in a finite number of sectors. Shukla, Hirsch and Schmeck propose in [135] to determine first an approximation (around 10,000 points) of the set of efficient elements w.r.t. the ordering cone $\mathbb{R}^m_+$ using classical multiobjective evolutionary algorithms, and then to select among the approximation points the equitable efficient points.

Another approach for solving vector optimization problems with a variable ordering structure is presented by Bello Cruz and Bouza Allende in [14]: for determining minimal elements a steepest descent-like method for smooth unconstrained vector optimization problems is developed. The procedure does not rely on a scalarization. Convergence to a minimal element under convexity assumptions is proven.

The numerical approaches presented in this chapter were first presented in [49]. Many numerical experiments with the methods discussed here can be found in the diploma theses of Ziegler [162].

Chapter 10
Outlook and Further Application Areas

In this final chapter we discuss some additional applications and related mathematical areas in which variable ordering structures appear. While the relation between these other areas and the results presented in this book are not yet fully explored, the application in intensity-modulated radiation therapy motivates to study also more general concepts of variable ordering structures.

As we have seen in Sect. 1.3, variable ordering structures appear in vector optimization problems, for instance, when modeling problems from medical image registration or portfolio optimization. Closely related to the study of vector optimization problems are vector variational inequalities and there, variable ordering structures are also studied. In Sect. 10.1 we shortly present in which form variable ordering structures appear in that context. In Sect. 10.2 we recall the concept of local preferences as discussed in the theory of consumer demand and point out their relation to variable ordering structures. Finally, we end this chapter by another application problem: in intensity-modulated radiation therapy the problem of finding an optimal treatment plan is modeled so far as a multiobjective optimization problem with the objective space partially ordered by the componentwise (natural) partial ordering. We discuss the limitations of this model and show that variable ordering structures have to be considered. We also point out that variable ordering structures based on a cone-valued ordering map may only locally be the right concept.

10.1 Vector Variational Inequalities and Related Problems

Vector variational inequalities and their relation to vector optimization problems in partially ordered spaces have been intensively studied in the last decades since their introduction by Giannessi in 1980 [67]. Thus it is a natural consequence that also vector variational inequalities with variable ordering cones are studied. In [28], Chen examines the following vector variational inequality (VVI): Let X, Y be real Banach spaces, $S \subset X$ a nonempty closed convex set and $T: S \to \mathcal{L}(X, Y)$ a map with $\mathcal{L}(X, Y)$ the space of continuous linear maps from X into Y. Additionally, we

G. Eichfelder, *Variable Ordering Structures in Vector Optimization*, Vector Optimization,   175
DOI 10.1007/978-3-642-54283-1_10, © Springer-Verlag Berlin Heidelberg 2014

assume a set-valued map $C: S \to 2^Y$ to be given with $C(x)$ a closed pointed convex cone with nonempty interior for all $x \in S$. The task, which is related to optimality conditions in mathematical optimization, is now to find some $\bar{x} \in S$ such that

$$T(\bar{x})(x - \bar{x}) \notin -\text{int}(C(\bar{x})) \quad \text{for all } x \in S. \tag{10.1}$$

In [107], Lee, Kim and Kuk use in this context the notion of a *generalized efficient solution* of a vector optimization problem $\min_{x \in S} f(x)$, which is similar to the notion of a weakly minimal solution given in Definition 2.33. They consider a vector-valued objective map f and assume the map C to be defined as above. An element $\bar{x} \in S$ is called a generalized efficient solution if there is no $x \in S$ such that

$$f(x) \in \{f(\bar{x})\} - \text{int}(C(\bar{x})).$$

This notion was also used by Ansari and Yao in [6] setting $S = X$ and calling it (global) vector minimum point. In [68], Giannessi, Mastroeni and Yang use the name weakly minimal solution. Clearly, if f is a bijective single-valued map, we can define the cone-valued map $\mathcal{D}$ on the image set by $\mathcal{D}(y) := C(x)$ with x given by the equation $y = f(x)$. Then $\bar{x}$ is a generalized efficient solution w.r.t. the ordering map C if and only if $(\bar{x}, f(\bar{x}))$ is a weakly minimal solution w.r.t. the ordering map $\mathcal{D}$ in the sense of Definition 2.33.

Several extensions of the above VVI are examined for instance by Zheng [161] and Ceng and Huang [26], see also the book by Göpfert et al. [71] and the references therein. In this context also a nonlinear scalarization functional is used to define a so-called gap function, see e.g. [3] by Al-Homodan, Ansari and Schaible. Let $r(x) \in \text{int}(C(x))$ be given for all $x \in S$, then Chen [30, 31] defines the functional $\xi_r : S \times Y \to \mathbb{R}$ by

$$(x, y) \mapsto \inf\{t \in \mathbb{R} \mid y \in \{t \cdot r(x)\} - C(x)\} \quad \text{for all } x \in S, \ y \in Y.$$

This scalarization is obviously closely related to the nonlinear scalarization which we have discussed in Sect. 5.2, see also the notes on the literature in Sect. 5.3.

Variable ordering structures are also presumed in the context of vector complementarity problems, see for instance [87] by Huang, Yang and Chan or the book section [29, Sect. 3.9] by the same authors: with T and C as above, setting $S := X$, and assuming that $K \subset Y$ is a convex cone, they examine the following weak vector complementarity problem: Find some $\bar{x} \in K$ such that

$$T(\bar{x})(\bar{x}) \notin \text{int}(C(\bar{x})) \ \wedge \ T(\bar{x})(x) \notin -\text{int}(C(\bar{x})) \quad \text{for all } x \in K.$$

Giannessi, Mastroeni and Yang list in [68] that these vector complementarity problems with a variable ordering structure have applications in the analysis of dynamic problems as time dependent traffic equilibrium problems, referring to Daniele et al.

[35] and Khanh and Luu [101]. Relations to special vector optimization problems can be examined using the notion of a generalized efficient solution as given above, as well as to special vector variational inequalities.

Finally, also in vector (quasi-)equilibrium problems families of convex cones are of interest. These problems allow a unified treatment of several topics in optimization as vector optimization problems and vector variational inequalities, see, for instance, Lee et al. [108], Li and Li [109] and Mastroeni [115] and the references therein: for a vector-valued map $f: X \times X \to Y$ with $f(x, x) = 0$ for all $x \in X$ and a set-valued map $K: X \to 2^X$ some $\bar{x} \in K(\bar{x})$ is searched with

$$f(x, \bar{x}) \notin C(\bar{x}) \setminus \{0_Y\}.$$

10.2 Theory of Consumer Demand in Economics

In the traditional theory of consumer demand, see [97,98] by John and the references therein, one assumes that a consumer's choice is derived from maximizing the utility. It is differentiated between the local and the global preferences, where the local preferences in the space $Y = \mathbb{R}^m$ are expressed by the following: Let $y \in \mathbb{R}^m$ be given. A direction $d \in \mathbb{R}^m$ is preferred if

$$w(y)^\top d < 0,$$

non-preferred if $w(y)^\top d > 0$, and indifferent if $w(y)^\top d = 0$ with $w: \mathbb{R}^m \to \mathbb{R}^m$ some function. We present the concepts of [98] using the notations of this book: to any element $y \in \mathbb{R}^m$ the cone of dominated or non-preferred elements, including the indifferent elements, is given by

$$\mathcal{D}(y) = \{d \in \mathbb{R}^m \mid w(y)^\top d \geq 0\},$$

which is a convex cone which is not pointed. In fact, $\mathcal{D}(y)$ is a halfspace.

According to Allen [4] and Georgescu-Roegen [62,63] a point $\bar{y} \in \mathbb{R}^m$ is defined to be an equilibrium position, if no direction away from $\bar{y}$ to any other alternative y is preferred, i.e. if

$$w(\bar{y})^\top (y - \bar{y}) \geq 0.$$

This corresponds to $y \in \{\bar{y}\} + \mathcal{D}(\bar{y})$ for all feasible y, i.e. $\bar{y}$ has to be a strongly minimal element according to Definition 2.12(d) (where the definition can also be used for $\mathcal{D}(y)$ being a non-pointed cone as it is the case here).

To guarantee additionally the equilibrium to be stable, the so-called principle of persisting nonpreferences, it is required, see [63, 98], that for any $y \in \mathbb{R}^m$ it holds that

$$w(y)^\top d \geq 0 \text{ implies } w(y + d)^\top d \geq 0.$$

Equivalently, this can be written as

$$d \in \mathcal{D}(y + d) \text{ for all } d \in \mathcal{D}(y).$$

This property corresponds to the map w being pseudomonotone. We assumed the same condition in (2.12) in Lemma 2.50. Additionally, w is assumed to be continuous and is then called a local preference representation on Y.

10.3 Variable Preferences in Intensity-Modulated Radiation Therapy

In medical engineering, to be more concrete in intensity-modulated radiation therapy (IMRT), Thieke suggested recently (C. Thieke, 2010, private communication) the incorporation of variable ordering structures to allow an improved modeling of the decision making problem. In IMRT one searches for an optimal treatment plan for the irradiation of a tumor with the target to spare the surrounding tissue, or at least to reduce the radiation dose delivered to the neighbored healthy organs, while destroying the tumor.

This problem is modeled so far as a multiobjective optimization problem with an objective for each healthy neighbored organ measuring its dose stress. For instance, for the treatment of a prostate cancer tumor, the two objective functions, which have to be minimized, are the dose delivered to the rectum (organ A) and the dose delivered to the bladder (organ B). For comparing different treatment plans the natural ordering in the objective space, i.e. the componentwise partial ordering, is used so far. Hence, the treatment plans with the image values marked as dots in Fig. 10.1 are not comparable, as no reduction in the dose delivered to one organ is possible without delivering more dose to the other organ.

Having a look on the dose–response curve shows that this does not appropriately model the problem. A dose–response curve describes thereby the caused effect to a particular volume (organ) of interest according to increasing dose, as illustrated in Fig. 10.2. As long as the response of the organ on dose variations is relatively small, which corresponds to the flat parts in the dose–response curve, a rise of the dose delivered to that organ in favor of an improvement of the value for another organ is preferred. For instance, if we have the dose A_1 delivered to the rectum and B_1 delivered to the bladder, we would not only prefer an improvement of the dose level in both organs but also to rise the dose delivered to the rectum to the level A_2 for reducing the dose amount in the bladder, for instance, to B_2. The reason for that is that a rise in the dose level for the rectum to A_2 has hardly any influence on the effect in that organ while for the bladder a large improvement in the effect on that organ is reached by changing to B_2. Thus, dependent on the actual values in the objective space, different preferred directions exist. For (A_1, B_1), $d := (A_2 - A_1, B_2 - B_1)$ is a preferred direction.

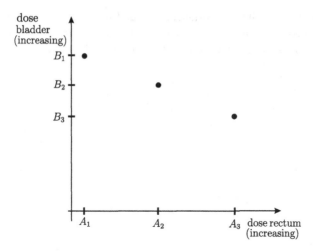

Fig. 10.1 Different achievable values for the dose delivered to the rectum and to the bladder for three different treatment plans

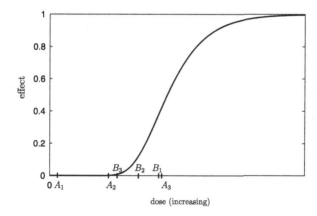

Fig. 10.2 Dose–response curve: portion of effect according to increasing dose level. We assume the same curve for both organs, the rectum and the bladder, cf. [47]

However, if we model these sets of preferred directions by a cone-valued ordering map, i.e. by denoting any positive multiplier λd of d with $\lambda > 0$ also preferred, then this implies that also a rise from A_1 to A_3 in favor of a reduction from B_1 to B_3 in the levels for the rectum and the bladder, respectively, would be preferred. But as we reach now for the dose level of the rectum the sensible (steep) part in the dose–response curve, and the increase in the effect for the rectum is large, this might not be a preferred solution from a practical point of view. Hence, a cone-valued ordering map for describing the variable ordering structure is here only locally adequate. For a global view on the problem other maps $\mathcal{D}$ for describing the set of preferred elements $\{y\} - \mathcal{D}(y)$ (dominated elements $\{y\} + \mathcal{D}(y)$, respectively) for some element y with, for instance, bounded image sets or image sets being the union of a convex cone and a bounded set have to be considered. For a study of domination sets instead of cones see Bergstresser et al. [19], Chew [32] and Weidner [151, 153].

References

1. M. Adán, V. Novo, Partial and generalized subconvexity in vector optimization problems. J. Convex Anal. **8**, 583–594 (2001)
2. M. Adán, V. Novo, Optimality conditions for vector optimization problems with generalized convexity in real linear spaces. Optimization **51**, 73–91 (2002)
3. S. Al-Homodan, Q.H. Ansari, S. Schaible, Existence of solutions of systems of generalized implicit vector variational inequalities. J. Optim. Theory Appl. **134**, 515–531 (2007)
4. R.G.D. Allen, The foundations of a mathematical theory of exchange. Economica **12**, 197–226 (1932)
5. G.B. Allende, C. Tammer, Scalar functions for computing minimizers under variable order structures, in *Congreso Latino-Iberoamericano de Investigación Operativa*, Rio de Janeiro, 2012
6. Q.H. Ansari, J.C. Yao, On nondifferentiable and nonconvex vector optimization problems. J. Optim. Theory Appl. **106**, 475–488 (2000)
7. J.P. Aubin, I. Ekeland, *Applied Nonlinear Analysis* (Wiley, New York, 1984)
8. J.P. Aubin, H. Frankowska, *Set-Valued Analysis* (Birkäuser, Boston, 1990)
9. D. Baatar, M.M. Wiecek, Advancing equitability in multiobjective programming. Comput. Math. Appl. **52**, 225–234 (2006)
10. B. Bank, J. Guddat, D. Klatte, B. Kummer, K. Tammer, *Non-Linear Parametric Optimization* (Akademie-Verlag, Berlin, 1982)
11. T.Q. Bao, B.S. Mordukhovich, Necessary nondomination conditions in set and vector optimization with variable ordering structures. J. Optim. Theory Appl. (2013). doi:10.2007/s10957-013-0332-6
12. E.M. Bednarczuk, Bishop-Phelps cones and convexity: applications to stability of vector optimization problems. INRIA Rapport de Recherche, No. 2806 (1996)
13. E.M. Bednarczuk, M.J. Przybyla, The vector-valued variational principle in Banach spaces ordered by cones with nonempty interior. SIAM J. Optim. **18**, 907–913 (2007)
14. J.Y. Bello Cruz, G. Bouza Allende, A steepest descent-like method for variable order vector optimization problems. J. Optim. Theory Appl. (2013). doi:10.1007/s10957-013-0308-6
15. J. Benoist, N. Popovici, Characterizations of convex and quasiconvex set-valued maps. Math. Method Oper. Res. **57**, 427–435 (2003)
16. H.P. Benson, An improved definition of proper efficiency for vector maximization with respect to cones. J. Math. Appl. **71**, 232–241 (1979)
17. C. Berge, *Topological Spaces* (Oliver and Boyd, Edinburgh, 1963)
18. C. Berge, *Espaces topologiques, Fonctions multivoques* (Dunod, Paris, 1966)
19. K. Bergstresser, A. Charnes, P.L. Yu, Generalization of domination structures and nondominated solutions in multicriteria decision making. J. Optim. Theory Appl. **18**, 3–13 (1976)

20. K. Bergstresser, P.L. Yu, Domination structures and multicriteria problems in N-person games. Theory Decis. **8**, 5–48 (1977)
21. E. Bishop, R.R. Phelps, The support functionals of a convex set. Proc. Symp. Pure Math. **7**, 27–35 (1962)
22. J. Borde, J.P. Crouzeix, Continuity properties of the normal cone to the level sets of a quasiconvex function. J. Optim. Theory Appl. **66**, 415–429 (1990)
23. J.M. Borwein, Proper efficient points for maximizations with respect to cones. SIAM J. Control Optim. **15**, 57–63 (1977)
24. J.M. Borwein, The geometry of Pareto efficiency over cones. Optimization **11**, 235–248 (1980)
25. R.I. Boţ, S.-M. Grad, G. Wanka, *Duality in Vector Optimization* (Springer, Heidelberg, 2009)
26. L.-C. Ceng, S. Huang, Existence theorems for generalized vector variational inequalities with a variable ordering relation. J. Glob. Optim. **46**, 521–535 (2010)
27. L.C. Ceng, B.S. Mordukhovich, J.C. Yao, Hybrid approximate proximal method with auxiliary variational inequalities for vector optimization. J. Optim. Theory Appl. **146**, 267–303 (2010)
28. G.Y. Chen, Existence of solutions for a vector variational inequality: an extension of the Hartmann-Stampacchia theorem. J. Optim. Theory Appl. **74**, 445–456 (1992)
29. G.Y. Chen, X. Huang, X. Yang, *Vector Optimization, Set-Valued and Variational Analysis* (Springer, Berlin, 2005)
30. G.Y. Chen, X.Q. Yang, Characterizations of variable domination structures via nonlinear scalarization. J. Optim. Theory Appl. **112**, 97–110 (2002)
31. G.Y. Chen, X.Q. Yang, H. Yu, A nonlinear scalarization function and generalized quasi-vector equilibrium problems. J. Glob. Optim. **32**, 451–466 (2005)
32. K.L. Chew, Domination structures in abstract spaces, in *Southeast Asian Bulletin of Mathematics*. Proceedings of the First Franco-Southeast Asian Mathematical Conference, 1979, pp. 190–204
33. F.H. Clarke, Optimal solutions to differential inclusions. J. Optim. Theory Appl. **19**, 469–479 (1976)
34. F.H. Clarke, *Optimization and Nonsmooth Analysis* (SIAM, New York, 1990). Reprint, originally published: Wiley, New York, 1983
35. P. Daniele, A. Maugeri, W. Oettli, Time-dependent variational inequalities. J. Optim. Theory Appl. **103**, 543–555 (1999)
36. J.P. Dauer, R.J. Gallagher, Positive proper efficient points and related cone results in vector optimization theory. SIAM J. Control Optim. **28**, 158–172 (1990)
37. J.P. Delahaye, J. Denel, Annex 1: the continuities of the point-to-set maps, definitions and equivalences. Math. Program. Study **10**, 8–12 (1979)
38. F.Y. Edgeworth, *Mathematical Psychics* (Kegan Paul, London, 1881)
39. M. Ehrgott, *Multicriteria Optimization* (Springer, Heidelberg, 2005)
40. G. Eichfelder, Parametergesteuerte Lösung nichtlinearer multikriterieller Optimierungsprobleme. Dissertation, Universität Erlangen-Nürnberg, 2006
41. G. Eichfelder, ε-Constraint method with adaptive parameter control and an application to intensity-modulated radiation therapy, in *Multicriteria Decision Making and Fuzzy Systems, Theory, Methods and Applications*, ed. by K.-H. Küfer et al. (Shaker, Aachen, 2006), pp. 25–42
42. G. Eichfelder, *Adaptive Scalarization Methods in Multiobjective Optimization* (Springer, Heidelberg, 2008)
43. G. Eichfelder, An adaptive scalarization method in multi-objective optimization. SIAM J. Optim. **19**, 1694–1718 (2009)
44. G. Eichfelder, Scalarizations for adaptively solving multi-objective optimization problems. Comput. Optim. Appl. **44**, 249–273 (2009)
45. G. Eichfelder, A constraint method in nonlinear multi-objective optimization, in *Multiobjective Programming and Goal Programming, Theoretical Results and Practical Applications*,

ed. by V. Barichard et al. Lecture Notes in Economics and Mathematical Systems, vol. 618 (Springer, Heidelberg, 2009), pp. 3–12

46. G. Eichfelder, Optimal elements in vector optimization with a variable ordering structure. J. Optim. Theory Appl. **151**(2), 217–240 (2011)

47. G. Eichfelder, Variable ordering structures in vector optimization, in *Recent Developments in Vector Optimization*, Chap. 4, ed. by Q.H. Ansari, J.-C. Yao (Springer, Heidelberg, 2012), pp. 95–126

48. G. Eichfelder, Cone-valued maps in optimization. Appl. Anal. **91**(10), 1831–1846 (2012)

49. G. Eichfelder, Numerical procedures in multiobjective optimization with variable ordering structures. J. Optim. Theory Appl. (2013). doi:10.1007/s10957-013-0267-y

50. G. Eichfelder, Ordering structures in vector optimization and applications in medical engineering. Preprint-Series of the Institute of Mathematics, Technische Universität Ilmenau, 2013

51. G. Eichfelder, M. Gebhardt, Local specific absorption rate control for parallel transmission by virtual observation points. Magn. Reson. Med. **66**(5), 1468–1476 (2011)

52. G. Eichfelder, T.X.D. Ha, Optimality conditions for vector optimization problems with variable ordering structures. Optimization **62**(5), 597–627 (2013)

53. G. Eichfelder, J. Jahn, Vector optimization problems and their solution concepts, in *Recent Developments in Vector Optimization*, Chap. 1, ed. by Q.H. Ansari, J.-C. Yao (Springer, Heidelberg, 2012), pp. 1–27

54. G. Eichfelder, R. Kasimbeyli, Properly optimal elements in vector optimization with variable ordering structures. J. Glob. Optim. (2013). doi:10.1007/s10898-013-0132-4

55. G. Eichfelder, T. Gerlach, Characterization of proper optimal elements with variable ordering structures. Preprint-Series of the Institute of Mathematics, Ilmenau University of Technology, Germany, 2014, http://www.db-thueringen.de/servlets/DerivateServlet/Derivate-28793/IfM_Preprint_M_14_01.pdf

56. B. El Abdouni, L. Thibault, Optimality conditions for problems with set-valued objectives. J. Appl. Anal. **2**, 183–201 (1996)

57. A. Engau, Domination and decomposition in multiobjective programming. Dissertation, University of Clemson, 2007

58. A. Engau, Variable preference modeling with ideal-symmetric convex cones. J. Glob. Optim. **42**, 295–311 (2008)

59. K. Fan, Minimax theorems. Proc. Natl. Acad. Sci. USA **39**, 42–47 (1953)

60. B. Fischer, E. Haber, J. Modersitzki, Mathematics meets medicine - an optimal alignment. SIAG/OPT Views News **19**(2), 1–7 (2008)

61. C. Gebhardt, Skalarisierungen für Vektoroptimierungsprobleme mit variablen Ordnungsstrukturen. Diploma thesis, Universität Erlangen-Nürnberg, 2011

62. N. Georgescu, The pure theory of consumer's behaviour. Q. J. Econ. **50**, 545–593 (1936)

63. N. Georgescu, Choice and revealed preference. South. Econ. J. **21**, 119–130 (1954)

64. C. Gerstewitz (Tammer), Nichtkonvexe Dualität in der Vektoroptimierung. Wiss. Z. TH Leuna-Merseburg **25**, 357–364 (1983)

65. C. Gerstewitz (Tammer), E. Iwanow, Dualität für nichtkonvexe Vektoroptimierungsprobleme. Wiss. Z. Techn. Hochschule Ilmenau **31**, 61–81 (1985)

66. C. Gerth (Tammer), P. Weidner, Nonconvex separation theorems and some applications in vector optimization. J. Optim. Theory Appl. **67**, 297–320 (1990)

67. F. Giannessi, Theorems of the alternative, quadratic programs and complementarity problems, in *Variational Inequalities and Complementarity Problems*, ed. by R.W. Cottle et al. (Wiley, New York, 1980)

68. F. Giannessi, G. Mastroeni, X.Q. Yang, Survey on vector complementarity problems. J. Optim. Theory Appl. **53**(1), 53–67 (2012)

69. G. Godini, Set-valued Cauchy functional equation. Rev. Roumaine Math. Pures Appl. **20**, 1113–1121 (1975)

70. A. Göpfert, R. Nehse, *Vektoroptimierung: Theorie, Verfahren und Anwendungen* (Teubner, Leipzig, 1990)

71. A. Göpfert, H. Riahi, C. Tammer, C. Zălinescu, *Variational Methods in Partially Ordered Spaces* (Springer, New York, 2003)
72. D. Gourion, D.T. Luc, Generating the weakly efficient set of nonconvex multiobjective problems. J. Glob. Optim. **41**, 517–538 (2008)
73. P.J. Guerra, M.A. Melguizo, M.J. Muñoz-Bouzo, Conic set-valued maps in vector optimization. Set-Valued Anal. **15**, 47–59 (2007)
74. P.J. Guerra, M.A. Melguizo, M.J. Muñoz-Bouzo, Polar conic set-valued map in vector optimization. Continuity and derivability. J. Optim. Theory Appl. **142**, 343–354 (2009)
75. A. Guerraggio, E. Molho, A. Zaffaroni, On the notion of proper efficiency in vector optimization. J. Optim. Theory Appl. **82**, 1–21 (1994)
76. C. Gutiérrez, B. Jiménez, V. Novo, On approximate solutions in vector optimization problems via scalarization. Comput. Optim. Appl. **35**, 305–324 (2006)
77. T.X.D. Ha, Optimality conditions for several types of efficient solutions of set-valued optimization problems, in *Nonlinear Analysis and Variational Problems*, Chap. 21, ed. by P. Pardalos, Th.M. Rassias, A.A. Khan (Springer, Heidelberg, 2009), pp. 305–324
78. T.X.D. Ha, Optimality conditions for various efficient solutions involving coderivatives: from set-valued optimization problems to set-valued equilibrium problems. Nonlinear Anal. **75**(3), 1305–1323 (2012)
79. T.X.D. Ha, J. Jahn, Properties of Bishop-Phelps cones. Preprint series of the Institute of Applied Mathematics, Universität Erlangen-Nürnberg, No. 343, 2010
80. A. Hamel, Translative sets and functions and their applications to risk measure theory and nonlinear separation. Preprint series of IMPA, Rio de Janeiro, 21, 2006
81. S. Helbig, An interactive algorithm for nonlinear vector optimization. Appl. Math. Optim. **22**(2), 147–151 (1990)
82. S. Helbig, Approximation of the efficient point set by perturbation of the ordering cone. Z. Oper. Res. **35**(3), 197–220 (1991)
83. I. Henig, Proper efficiency with respect to cones. J. Optim. Theory Appl. **36**, 387–407 (1982)
84. J.-B. Hiriart-Urruty, New concepts in nondifferentiable programming. Bull. Soc. Math. France **60**, 57–85 (1979)
85. J.-B. Hiriart-Urruty, Tangent cones, generalized gradients and mathematical programming in Banach spaces. Math. Oper. Res. **4**, 79–97 (1979)
86. C. Hirsch, P.K. Shukla, H. Schmeck, Variable preference modeling using multi-objective evolutionary algorithms, in *Evolutionary Multi-criterion Optimization - 6th International Conference*, ed. by R.H.C. Takahashi et al. Lecture Notes in Computer Science, vol. 6576 (Springer, Heidelberg, 2011)
87. N.J. Huang, X.Q. Yang, W.K. Chan, Vector complementarity problems with a variable ordering relation. Eur. J. Oper. Res. **176**, 15–26 (2007)
88. A.D. Ioffe, Approximate subdifferentials and applications 3: the metric theory. Mathematika **36**, 1–38 (1989)
89. A.D. Ioffe, J.-P. Penot, Subdifferentials of performance functions and calculus of coderivatives of set-valued mappings. Well-posedness and stability of variational problems. Serdica Math. J. **22**, 257–282 (1996)
90. G. Isac, Sur l'existence de l'optimum de Pareto. Riv. Mat. Univ. Parma, IV. Ser. **9**, 303–325 (1983)
91. G. Isac, A.O. Bahya, Full nuclear cones associated to a normal cone. Application to Pareto efficiency. Appl. Math. Lett. **15**, 633–639 (2002)
92. J. Jahn, *Introduction to the Theory of Nonlinear Optimization*, 3rd edn. (Springer, Heidelberg, 2007)
93. J. Jahn, Bishop-Phelps cones in optimization. Int. J. Optim. Theory Methods Appl. **1**, 123–139 (2009)
94. J. Jahn, *Vector Optimization - Theory, Applications, and Extensions*, 2nd edn. (Springer, Heidelberg, 2011)
95. J. Jahn, T.X.D. Ha, New order relations in set optimization. J. Optim. Theory Appl. **148**, 209–236 (2011)

96. J. Jahn, U. Rathje, Graef-Younes method with backward iteration, in *Multicriteria Decision Making and Fuzzy Systems - Theory, Methods and Applications*, ed. by K.-H. Küfer et al. (Shaker, Aachen, 2006), pp. 75–81

97. R. John, The concave nontransitive consumer. J. Glob. Optim. **20**, 297–308 (2001)

98. R. John, Local and global consumer preferences, in *Generalized Convexity and Related Topics*, ed. by I. Konnov, D.T. Luc, A. Rubinov (Springer, Heidelberg, 2006), pp. 315–326

99. E.K. Karaskal, W. Michalowski, Incorporating wealth information into a multiple criteria decision making model. Eur. J. Oper. Res. **150**, 204–219 (2003)

100. R. Kasimbeyli, A nonlinear cone separation theorem and scalarization in nonconvex vector optimization. SIAM J. Optim. **20**, 1591–1619 (2010)

101. P.Q. Khanh, L.M. Luu, On the existence of solutions to vector quasi-variational inequalities and quasi-complementarity problems with applications to traffic equilibria. J. Optim. Theory Appl. **123**, 533–548 (2004)

102. P. Korhonen, J. Wallenius, S. Zionts, Solving the discrete multiple criteria problem using convex cones. Manag. Sci. **30**, 1336–1345 (1984)

103. M.M. Kostreva, W. Ogryczak, A. Wierzbicki, Equitable aggregations in multiple criteria analysis. Eur. J. Oper. Res. **158**, 362–377 (2004)

104. M.A. Krasnosel'skij, *Positive Solutions of Operator Equations* (Noordhoff, Groningen, 1964)

105. D. Kuroiwa, Convexity for set-valued maps. Appl. Math. Lett. **9**, 97–101 (1996)

106. D. Kuroiwa, Natural criteria of set-valued optimization. Manuscript, Shimane University, 1998

107. G.M. Lee, D.S. Kim, H. Kuk, Existence of solutions for vector optimization problems. J. Math. Anal. Appl. **220**, 90–98 (1998)

108. G.M. Lee, D.S. Kim, B.S. Lee, On noncooperative vector equilibrium. Indian J. Pure Appl. Math. **27**, 735–739 (1996)

109. S.J. Li, M.H. Li, Levitin-Polyak well-posedness of vector equilibrium problems. Math. Methods Oper. Res. **69**, 125–140 (2009)

110. C.G. Liu, K.F. Ng, W.H. Yang, Merit functions in vector optimization. Math. Program. Ser. A **119**, 215–237 (2009)

111. J. Liu, W. Sond, On proper efficiencies in locally convex spaces—a survey. Acta Math. Vietnam. **26**(3), 301–312 (2001)

112. D.T. Luc, Scalarization of vector optimization problems. J. Optim. Theory Appl. **55**, 85–102 (1987)

113. D.T. Luc, *Theory of Vector Optimization* (Springer, Berlin, 1989)

114. D.T. Luc, J.P. Penot, Convergence of asymptotic directions. Trans. Am. Math. Soc. **353**, 4095–4121 (2001)

115. G. Mastroeni, On the image space analysis for vector quasi-equilibrium problems with a variable ordering relation. J. Glob. Optim. **53**(2), 203–214 (2012)

116. K.M. Miettinen, *Nonlinear Multiobjective Optimization* (Kluwer, Boston, 1999)

117. B.S. Mordukhovich, Maximum principle in problems of time optimal control with nonsmooth constraints. J. Appl. Math. Mech. **40**, 960–969 (1976)

118. B.S. Mordukhovich, Metric approximations and necessary optimality conditions for general classes of nonsmooth extremal problems. Sov. Math. Dokl. **22**, 526–530 (1980)

119. B.S. Mordukhovich, *Variational Analysis and Generalized Differentiation, I: Basic Theory, Grundlehren Series (Fundamental Principles of Mathematical Sciences)*, vol. 330 (Springer, Berlin, 2006)

120. B.S. Mordukhovich, *Variational Analysis and Generalized Differentiation, II: Applications, Grundlehren Series (Fundamental Principles of Mathematical Sciences)*, vol. 331 (Springer, Berlin, 2006)

121. B.S. Mordukhovich, Multiobjective optimization problems with equilibrium constraints. Math. Program. Ser. B **117**, 331–354 (2009)

122. B.S. Mordukhovich, Methods of variational analysis in multiobjective optimization. Optimization **58**, 413–430 (2009)

123. K. Nikodem, D. Popa, On single-valuedness of set-valued maps satisfying linear inclusions. Banach J. Math. Anal. **3**, 44–51 (2009)
124. Z.G. Nishnianidze, Fixed points of monotonic multiple-valued operators. Bull. Georgian Acad. Sci. **114**, 489–491 (1984)
125. W. Ogryczak, T. Sliwinski, On solving linear programs with the ordered weighted averaging objective. Eur. J. Oper. Res. **148**, 80–91 (2003)
126. W. Ogryczak, A. Wierzbicki, On multi-criteria approaches to bandwith allocation. Control Cybern. **33**, 427–448 (2004)
127. V. Pareto, *Manuale di economia politica* (Societa Editrice Libraria, Milano, 1906). (English translation: V. Pareto, *Manual of Political Economy*, translated by A.S. Schwier, M. Augustus, Kelley, New York, 1971)
128. A. Pascoletti, P. Serafini, Scalarizing vector optimization problems. J. Optim. Theory Appl. **42**, 499–524 (1984)
129. M. Petschke, On a theorem of Arrow, Barankin, and Blackwell. SIAM J. Control Optim. **28**, 395–401 (1990)
130. M. Pruckner, Kegelwertige Abbildungen in der Optimierung. Diploma thesis, Universität Erlangen-Nürnberg, 2011
131. J.H. Qiu, Y. Hao, Scalarization of Henig properly efficient points in locally convex spaces. J. Optim. Theory Appl. **147**, 71–92 (2010)
132. R. Ramesh, M.H. Karwan, S. Zionts, Preference structure representation using convex cones in multicriteria integer programming. Manag. Sci. **35**, 1092–1105 (1989)
133. A.M. Rubinov, R.N. Gasimov (Kasimbeyli), Scalarization and nonlinear scalar duality for vector optimization with preferences that are not necessarily a pre-order relation. J. Glob. Optim. **29**, 455–477 (2004)
134. Y. Sawaragi, H. Nakayama, T. Tanino, *Theory of Multiobjective Optimization* (Academic, London, 1985)
135. P.K. Shukla, C. Hirsch, H. Schmeck, In search of equitable solutions using multi-objective evolutionary algorithms, in *Parallel Problem Solving from Nature - PPSN XI*, ed. by R. Schaefer et al. Lecture Notes in Computer Science, vol. 6238 (Springer, Heidelberg, 2011), pp. 687–696
136. P.K. Shukla, M.A. Braun, Indicator based search in variable orderings: theory and algorithms, in *EMO 2013*, ed. by R.C. Purshouse et al. Lecture Notes in Computer Science, vol. 7811 (Springer, Heidelberg, 2013), pp. 66–80
137. W. Stadler, Fundamentals of multicriteria optimization, in *Multicriteria Optimization in Engineering and in the Sciences*, ed. by W. Stadler (Plenum, New York, 1988)
138. A. Sterna-Karwat, Continuous dependence of solutions on a parameter in a scalarization method. J. Optim. Theory Appl. **55**(3), 417–434 (1987)
139. A. Sterna-Karwat, Lipschitz and differentiable dependence of solutions on a parameter in a scalarization method. J. Aust. Math. Soc. A **42**, 353–364 (1987)
140. B. Soleimani, C. Tammer, Approximate solutions of vector optimization problem with variable ordering structure, in *AIP Conference Proceedings of Numerical Analysis and Applied Mathematics, ICNAAM 2012: International Conference of Numerical Analysis and Applied Mathematics*, vol. 1479, 2012, pp. 2363–2366
141. B. Soleimani, C. Tammer, Concepts for approximate solutions of vector optimization problems with variable order structure. Preprint-Series of the Institute of Mathematics, Martin-Luther-Universität Halle-Wittenberg, 2013
142. C. Tammer, C. Zălinescu, Lipschitz properties of the scalarization function and applications. Optimization **59**, 305–319 (2010)
143. C. Tammer, C. Zălinescu, Vector variational principles for set-valued functions, in *Recent Developments in Vector Optimization*, Chap. 11, ed. by Q.H. Ansari, J.-C. Yao (Springer, Heidelberg, 2011)
144. M. Tanaka, GA-based decision support system for multi-criteria optimization. Proc. Int. Conf. Syst. Man Cybern. **2**, 1556–1561 (1995)

145. L. Thibault, On subdifferentials of optimal value functions. SIAM J. Control Optim. **29**, 1019–1036 (1991)

146. M. Wacker, Multikriterielle Optimierung bei der Registrierung medizinischer Daten. Diploma thesis, Universität Erlangen-Nürnberg, 2008

147. M. Wacker, F. Deinzer, Automatic robust medical image registration using a new democratic vector optimization approach with multiple measures, in *Medical Image Computing and Computer-Assisted Intervention - MICCAI 2009*, ed. by G.-Z. Yang et al., 2009, pp. 590–597

148. D.W. Walkup, R.J.-B. Wets, Continuity of some convex-cone-valued maps. Proc. Am. Math. Soc. **18**, 229–235 (1967)

149. P. Weidner, Dominanzmengen und Optimalitätsbegriffe in der Vektoroptimierung. Wiss. Z. Techn. Hochschule Ilmenau **31**, 133–146 (1985)

150. P. Weidner, Problems in models and methods of vector optimization. Wiss. Schriftenr. TU Karl-Marx-Stadt **5**, 47–57 (1989)

151. P. Weidner, Ein Trennungskonzept und seine Anwendung auf Vektoroptimierungsverfahren. Habilitation thesis, Martin-Luther-Universität Halle-Wittenberg, 1990

152. P. Weidner, An approach to different scalarizations in vector optimization. Wiss. Z. Techn. Hochschule Ilmenau **36**, 103–110 (1990)

153. P. Weidner, Problems in scalarizing multicriteria approaches, in *Multiple Criteria Decision Making in the New Millennium*, ed. by M. Köksalen, S. Zionts (Springer, Heidelberg, 2001), pp. 199–209

154. M.M. Wiecek, Advances in cone-based preference modeling for decision making with multiple criteria. Decis. Mak. Manuf. Serv. **1**, 153–173 (2007)

155. G. Xiao, H. Xiao, S. Liu, Scalarization and pointwise well-posedness in vector optimization problems. J. Glob. Optim. **49**(4), 561–574 (2011)

156. Y.M. Younes, Studies on discrete vector optimization. Dissertation, University of Demiatta, 1993

157. R.C. Young, The algebra of many-valued quantities. Math. Ann. **104**, 260–290 (1931)

158. P.L. Yu, Cone convexity, cone extreme points, and nondominated solutions in decision problems with multiobjectives. J. Optim. Theory Appl. **14**, 319–377 (1974)

159. P.L. Yu, *Multiple-Criteria Decision Making: Concepts, Techniques and Extensions* (Plenum Press, New York, 1985)

160. A. Zaffaroni, Degrees of efficiency and degrees of minimality. SIAM J. Control Optim. **42**, 1071–1086 (2003)

161. F. Zheng, Vector variational inequalities with semi-monotone operators. J. Glob. Optim. **32**, 633–642 (2005)

162. M. Ziegler, Numerische Verfahren zur Bestimmung optimaler Elemente bei variablen Ordnungsstrukturen. Diploma thesis, Universität Erlangen-Nürnberg, 2012

Index

G. Eichfelder, *Variable Ordering Structures in Vector Optimization*, Vector Optimization,   189
DOI 10.1007/978-3-642-54283-1, © Springer-Verlag Berlin Heidelberg 2014

Printed in the United States
By Bookmasters